Life & Cognition at the Intersection of Science, Philosophy, & Religion

Life & Cognition at the Intersection of Science, Philosophy, & Religion

Science & Scientist 2023 Conference Proceedings

B Madhava Puri, PhD

Krishna Keshava Das

Bhakti Vedanta Institute
of Spiritual Culture and Science
Princeton, New Jersey, USA

Readers interested in the subject matter discussed in this book are encouraged to contact:

Dr. B. Mādhava Purī
princeton@bviscs.org
www.bviscs.org

Copyright © 2024 by Bhakti Vedanta Institute of Spiritual Culture and Science

All rights reserved. No part of this book may be reproduced in any manner whatsoever without written permission except in the case of brief quotations embodied in critical articles and reviews.

The book cover was created with the assistance of generative AI.

First Printing, 2024

Published by the Bhakti Vedanta Institute of Spiritual Culture and Science
Princeton, New Jersey, United States of America

Cataloging-in-Publication Data

Puri, Bhakti Madhava
Life & cognition at the intersection of science, philosophy, & religion: Science & scientist 2023 conference proceedings
Editorial advisor Bhakti Madhava Puri | editor Krishna Keshava Das
Includes bibliographical references
Print ISBN: 978-1-7349089-3-0
Ebook ISBN: 978-1-7349089-4-7

" It seems like it's a worthwhile organization asking important questions. "

Perry Marshall | Pioneer of $10M Evolution 2.0 Prize and invited speaker at Science & Scientist 2022

" I have had an opportunity to view the recording of the Life & Cognition symposium and want to congratulate the organizers and speakers for an outstanding intellectual accomplishment. "

James A Shapiro | Molecular Biologist and author of *Evolution: A View from the 21st Century. Fortified.* (2022)

CONTENTS

ABOUT THE CONFERENCE TOPIC xi

1 | Purposive Explanations Are More Useful For Identifying Lower-Level Activity In Living Systems Than The Other Way Round 1

2 | Superseding the Synaptic Network: How Cellular Complexity Transcends the Digital Neuron 7

3 | Death & Desire: Negativity at the Foundation of Life 12

4 | Understanding the Vedāntic View on the Difference between Life & Non-Life 19

5 | Emotional Intelligence in Cows 26

6 | The Problem of Embodied Consciousness in the Lens of Vedāntic View of Consciousness 31

CONTENTS

7 | Journey into Mind - DNA - Consciousness   35

8 | Evolution is Cognitive Thermodynamics   41

9 | Complexity Theory & Purposiveness   50

10 | Conclusion   61

REFERENCES   73
BVISCS PEER-REVIEWED PUBLICATIONS   75

These proceedings include a brief introduction of each speaker, the abstract that each speaker personally provided, and then a summary of each talk written by the editor.

Some summaries are followed by relevant notes that tie the talks together and bring them into a more cohesive dialogue with what the Princeton Bhakti Vedanta Institute hoped to discuss during Science & Scientist 2023: Life & Cognition at the Intersection of Science, Philosophy, & Religion.

ABOUT THE CONFERENCE TOPIC

A Historical Development of Life, Cognition, & Self in Modern Science

In 1980, biologist Humberto Maturana theorized that "[l]iving systems are cognitive systems and living as a process is a process of cognition. This statement is valid for all organisms with and without a nervous system." [1] Then in 1983, during her Nobel lecture, Barbara McClintock set a goal for 21st-century science to "determine the extent of knowledge the cell has of itself, and how it utilizes this knowledge in a 'thoughtful' manner when challenged." [2] In 2021, molecular biologist James A. Shapiro, a former student and colleague of McClintock, published the paper "All living cells are cognitive" in *Biochemical and Biophysical Research Communications*, thus empirically verifying Maturana's hypothesis and making progress towards McClintock's goal. [3] Recognizing that cognition/consciousness is ubiquitous throughout all lifeforms frustrates reductionist attempts to describe consciousness in terms of neuronal correlates — "the minimum neuronal mechanisms jointly sufficient for any one specific conscious experience" [4] — because (1) a single neuron, which is a single eukaryotic cell, already demonstrates cognitive behavior and (2) correlation does not imply causation. Comprehending the cause of cognition/consciousness requires starting from a foundation embracing all four aspects of Aristotelean causality — material, efficient, formal, and final — of which the formal and final aspects were abandoned after Francis Bacon (1561-1626). The purpose that something serves is the reason that it exists and that for-the-sake-of-which cognition/consciousness exists is the self.

McClintock's goal for modern science to determine "the extent of knowledge the cell has of itself" spurred enough momentum that self/nonself discrimination became subject to scientific analysis. Two aspects of self seem to be observed in cellular activity — universal and individual — such that (1) a collective of bacteria demonstrate discrimination between those belonging to their particular colony and those who do not and (2) individual bacterial cells distinguish between their own genetic material and foreign material. [5] It is worth noting that although cells are cognizant of various aspects of their existence, being self-conscious is a second-order awareness that requires being conscious of cognizant existence, which is distinct from being cognizant yet unaware of the ability. Cells do not demonstrate self-consciousness. Self-consciousness seems to be an accomplishment seen only in higher lifeforms like humans. Groundbreaking work on cellular cognition is relevant to cancer research, where cancer is sometimes viewed as an effect of cells within an organism becoming disoriented and resorting to a "unicellular lifestyle." [6] This scientific development has motivated other scientists to clarify the concept of "self" in a holistic manner.

Developmental/synthetic biologist Michael Levin proposes defining an individual Self by its information-processing and goal-seeking capacities, where smaller selves (like cells) and bigger selves (like organisms) are dynamically related and interdependent. He explains that the "computational boundary" of a self is the limit of its capacity to influence or achieve certain goals and that this conception is non-reductionistic due to the whole being greater than the sum of its parts because the goal being pursued by the whole (organism) is beyond the reach of each individual part/component (cell, or cellular collective like a tissue or organ) alone. He specifies that examples of self as he defines it concern "functional, third-person, objective capacities, computations, and behaviors." He intentionally avoids consciousness, which he distinguishes from cognition by defining consciousness as "first person experience or a sense of self as qualia." [7] But the perspective of first-person experience cannot be avoided when contemplating the self since that is the only concrete account we have of it, i.e. the direct experience of selfhood. All third-person perspectives are merely baseless speculation if not connected with our direct experience of selfhood. This tendency of scientists to avoid first-person perspectives is a byproduct of the historical development of modern science, where Francis Bacon deterred future generations of scientists from considering the first-person perspective by declaring "Of ourselves we say nothing" (Latin: *de nobis ipsis silemus*). [8] Thus Dr. B Mādhava Purī — the visionary behind this Science & Scientist conference series — recognizes that "science has lost its self-consciousness." [9]

Restoring self-consciousness to contemporary scientific investigation requires embracing a phenomenological approach — the "science of the experience of consciousness" as described by German philosopher G.W.F. Hegel. [10] Beyond his Phenomenology, Hegel explained that there is a range of categorical thought inherent to both our immediate external sensuous experience and mediated internal contemplative experience. The logical, natural, and spiritual categories deduced through conceptual thinking are not ego-centric speculation but the result of thought's own inner necessity and movement through negativity. Thought's movement proceeds forward, encounters opposition, progresses by reconciling opposition, and continues to develop in this manner, where "concepts indeed do give way necessarily to other succeeding concepts without which they would remain incomplete and to which they necessarily refer." [11] Within the scope of Hegel's methodology, the category of life arises in both Logic and Nature. Within Logic, life gives rise to the category of cognition, from which the Absolute Idea emerges. The transitions between logical life, cognition, and the Absolute Idea are not to be taken lightly and the development must be seriously considered.

Life & Cognition in Hegel & Vedānta

Hegel's philosophy has attracted the attention of humanities scholars at Princeton University several times since 2017, [12] [13] [14] including Alexander Englert, now a Research Associate at the Institute for Advanced Study in Princeton, who studies, among other topics, Hegel's logical category of life, its transition to the category of cognition, and the implications this has for comprehending organic life and artificial intelligence. [15] In summary, the transition from logical life to cognition, where the movement of pure logical thought remains free of sensuous limitations, involves:

1. Individual life — containing objectivity or the tendency to identify itself statically — absorbed in its immediate being-for-self indifferent to otherness, recognizes its identity with otherness as an extended aspect of its being-for-self because it is dependent on that otherness in that otherness determines the individual as being not other than itself. This serves to expand individual life's conception of wholeness beyond individuality to universality ("a 'big picture' conception of its [living] process as a whole" [16]).

ABOUT THE CONFERENCE TOPIC

2. The universal (genus) aspect was always implicitly in the background of individual life, unbeknownst to it, and upon being recognized by the individual, universality becomes explicit and negates the living individual, i.e. it becomes implicit in the universal genus as a sublated moment of a unified activity where "the most basic logical unit of the big picture receives conceptual determination as one of the many (logically similar) units in the same process," [17] serving to satisfy a sense of completion for the individual's being-for-self.

3. The next shift in this process is motivated by the reconciliation of the contradiction that arises when the individual recognizes that it is not merely for-itself, its sense of completion as a moment of the universal is not merely for the individual, but the universal is also for-itself, thus as a moment of the universal, the individual must be for the universal as well as for itself. The reconciliation of this contradiction reveals the dialectic interpenetrating relationship between the individual and universal, where the living individual realizes that its being-for-self is simultaneously identical with and different from the universal genus' being-for-self.

4. So the individual and universal form a negative unity such that when individual life is explicit the universal genus is implicit, and when the universal is explicit the individual is implicit. This dynamic, interpenetrating, dialectic, purely logical activity as a totality is a "processual whole," [18] i.e. the process itself is the whole and *not* any particular moment of it abstracted from the process and considered in static isolation.

5. When the objective tendency to identify with a particular static moment of the processual whole is fully eradicated and

ABOUT THE CONFERENCE TOPIC

replaced by complete absorption in the dynamic movement of the processual whole which is for itself, the transition to the logical category of cognition occurs.

"Embodied cognition" refers to the understanding that cognition/consciousness is enmeshed in the relationship between a body and interactions with its environment, as opposed to being isolated in the brain. In this regard, it is significant that although an individual's body is always changing — where all constitutive cells of a body [19] including neurons [20] are replaced several times throughout a single lifetime — we remain the same person. What does this imply for embodied cognition? Vedāntic knowledge explains that the self is the unchanging aspect that persists despite the constant change of the physical body, as described in *Bhagavad-gītā* 2.13:

> <u>dehino</u> 'smin yathā dehe   kaumāraṁ yauvanaṁ jarā
> tathā dehāntara-prāptir   dhīras tatra na muhyati

"As the <u>embodied soul</u> continuously passes, in this body, from boyhood to youth to old age, the soul similarly passes into another body at death. A sober person is not bewildered by such a change."

This verse explains that just as an individual soul/self remains the same self throughout the transitions of its aging material body, that same individual will also pass into a new body at the time of death (reincarnation). Here, the Sanskrit word *"dehino"* refers to the "embodied soul," where the constitution of the soul is *sat-cit-ānanda* — willing/volition, thinking/cognition/consciousness, and feeling/emotion. The materially embodied soul is afflicted by forgetfulness of its original spiritual nature in the plane of dedication, where its capacity for willing, thinking, and feeling are absorbed in a loving serving relation to Supreme Spirit. Thus, the materially embodied soul engages in exploitive activities focused on self-centered sense enjoyment, which gives rise to bodily, mental, and emotional suffering including the experience of death. This is due to exploitation being against the soul's intrinsically dedicative nature, like a fish out of water. When the embodied soul actively cultivates remembrance of its constitutional position, it begins transcending material suffering.

Vedānta describes that our consciousness determines our experience; our knowledge of the world determines the world we live in; our knowledge of an object determines what the object is. The position of embodied cognition coincides with Vedāntic knowledge in that cognition/consciousness is intertwined in the relationship between a body and its interactions with the environment, but understanding the body as a material thing intended to exploit the environment to maximize individual enjoyment produces a very different experience than identifying as spirit fixed in uninterrupted joyful loving service to the Supreme.

The Evolution of Consciousness

Clarifying the distinction between cognition and consciousness is useful for accurately determining the various behaviors of particular living entities. The word "cognition" comes from the Latin "*cognoscere*" which includes the root "*gno-*" indicating a general kind of knowing/knowledge as understood with the word "gnostic," [21] while the word "consciousness" stems from the Latin "*conscire*" which includes "*scire*" referring to a more expert kind of knowing as seen in the word "science." [22] So, the etymology seems to indicate

ABOUT THE CONFERENCE TOPIC

that the word cognition is intended to denote a less sophisticated faculty than the word consciousness.

Cognition is an initial step of mediated thought where an object is reflected into the mind and a preliminary mental representation is formed. This is sufficient to navigate the relatively simplistic experience of cellular life and other lower lifeforms such as insects and plants. The phrase "lower lifeforms" is not being used in a derogatory way but to denote living entities whose activities seem mainly absorbed in exercising volition and cognition while responding to immediate environmental circumstances without exhibiting symptoms of a sophisticated internal emotional experience. Consciousness, on the other hand, denotes an identity-in-difference between subject and object that requires a dialectic approach to comprehend. This is a further development of thought where the conscious agent knows apparently external objects as identical to itself, as well as recognizing the difference between itself and other objects. In addition to exercising volition and cognition, conscious entities like elephants, [23] cows, [24] and humans have a more mediated relationship with their environment allowing them to form emotional attachments to things other than themselves. Scientists recognize that "there is continuity between humans and other animals in their emotional (and cognitive) lives; that there are transitional stages among species, not large gaps; and that the differences among many animals are differences in degree rather than in kind." [25] When one experiences an identity with something other than oneself, a feeling of inner connection is established. So, living cells may be volitional and cognitive, but not emotional. This indicates an evolution of consciousness throughout lifeforms where different stages of development are distinguished by the degree to which and particular manner in which a living entity expresses volitional, cognitive, and emotional activity.

References

1. Maturana, Humberto. 1980. *Autopoiesis and Cognition: Realization of the Living,* 13. D. Reidel Publishing Company.

2. McClintock, Barbara. 1983. "The Significance of Responses of the Genome to Challenge," 193. Nobel Lecture. https://www.nobelprize.org/uploads/2018/06/mcclintock-lecture.pdf

3. Shapiro, James A. 2021. "All living cells are cognitive." *Biochemical and Biophysical Research Communications* 564. https://doi.org/10.1016/j.bbrc.2020.08.120

4. Koch, Christof, et al. 2016. "Neural correlates of consciousness: progress and problems," *Nature Reviews Neuroscience* 17. https://www.nature.com/articles/nrn.2016.22

5. Shapiro. "All living cells are cognitive," 3-7.

6. Marshall, Perry. 2021. "Biology transcends the limits of computation." *Progress in Biophysics and Molecular Biology* 165. https://doi.org/10.1016/j.pbiomolbio.2021.04.006

7. Levin, Michael. 2019. "The Computational Boundary of a 'Self': Developmental Bioelectricity Drives Multicellularity and Scale-Free Cognition." *Frontiers in Psychology* 10. https://www.frontiersin.org/articles/10.3389/fpsyg.2019.02688/full

8. "Francis Bacon, The Great Instauration (excerpts)" 337. Hanover Historical Texts Project. https://history.hanover.edu/texts/Bacon/gi.html

9. Purī, Bhakti Mādhava. 2020. *Idols of the Mind vs. True Reality* 79. Bhakti Vedanta Institute of Spiritual Culture and Science. ISBN 9781734908954.

10. Hegel, G.W.F. and A. V. Miller (trans).1977. *Phenomenology of Spirit* §88, p 56. Oxford University Press.

11. Englert, Alexander T. 2016. "Life, Logic, and the Pursuit of Purity: Logically Restructuring the Transition to Cognition." *Hegel Studien* 50. https://philarchive.org/rec/ENGLLA-2

12. "Hegel and the Humanities: A Symposium." September 2017. Princeton University Humanities Council. https://humanities.princeton.edu/event/hegel-and-the-humanities/

13. "Gauss Seminar Discusses the Importance of Hegel's Legacy." October 2017. Princeton University Humanities Council. https://humanities.princeton.edu/2017/10/05/gauss-seminar-discusses-the-importance-of-hegels-legacy/

14. "God & Infinity: Perspectives from Hegel and Kierkegaard." 2022. Princeton University Center for Human Values. https://uchv.princeton.edu/events/god-infinity-perspectives-hegel-and-kierkegaard

15. Englert, Alexander T. n.d. "Research." https://alexanderenglert.com/research/

16. Englert, "Life, Logic, and the Pursuit of Purity," 83.

17. Ibid 84.

18. Ibid 83.

19. Cole, Adam. 2016. "How do we know the lifespan of different tissue/cell types?" *NPR: Skunk Bear.* https://skunkbear.tumblr.com/post/146593746334/c14

20. Coffey, Donavyn. 2022. "Does the human body replace itself every 7 years?" Live Science. https://www.livescience.com/33179-does-human-body-replace-cells-seven-years.html

21. Douglas, Harper. 2023. "Etymology of cognition." Online Etymology Dictionary. https://www.etymonline.com/word/cognition

22. ———. 2018. "Etymology of consciousness." Online Etymology Dictionary. https://www.etymonline.com/word/consciousness

23. Bates, Lucy Anne. 2008. "Do Elephants Show Empathy?" *Journal of Consciousness Studies* 15(10). https://www.researchgate.net/publication/37245197_Do_Elephants_Show_Empathy

24. Marino, Lori, et al. 2017. "The Psychology of Cows." *Animal Behavior and Cognition* 4(4). https://dx.doi.org/10.26451/abc.04.04.06.2017

25. Bekoff, Marc. 2000. "Animal Emotions: Exploring Passionate Natures: Current interdisciplinary research provides compelling evidence that many animals experience such emotions as joy, fear, love, despair, and grief—we

are not alone." *Bioscience* 50(10). https://doi.org/10.1641/ 0006-3568(2000)050[0861:AEEPN]2.0.CO;2

Purposive Explanations Are More Useful For Identifying Lower-Level Activity In Living Systems Than The Other Way Round

Denis Noble, CBE, PhD, FRS | Oxford University

Latest book: *Understanding Living Systems (2023)*

About the Speaker: *Professor Emeritus and co-Director of Computational Physiology at Oxford University. One of the pioneers of Systems Biology and developed the first viable mathematical model of the working heart in 1960. Over 350 articles in academic journals, including Nature, Science, PNAS, Journal of Physiology.*

Abstract: *Physiology shows that higher-level functionality, including purposive explanations, are more successful in predicting lower-level, microscopic and molecular events than the other way round*

(Noble & Noble, 2023a, 2023b, 2023c). This is why GWAS association scores for most genes are abysmally low. So low, that they cannot be relied on to predict later life disease states (Hingorani et al, 2023). The reason is that low-level events are constrained by higher-level organisation by determining the boundary conditions under which those events operate.

References:
Hingorani, A.D. et al 2023. Performance of polygenic risk scores in screening, prediction, and risk stratification: secondary analysis of data in the Polygenic Score Catalog. British Medical Journal. e000554. doi:10.1136/ bmjmed-2023-000554

Noble R. & Noble, D. 2023a. Physiology restores purpose to evolutionary biology. Biological Journal of the Linnean Society. 139, 357-369.

Noble D. & Noble, R. 2023b. How purposive agency became banned from evolutionary biology. In Corning et al (Eds) Evolution 'on purpose'. MIT Press. 221-235.

Noble R. & Noble, D. 2023c. Understanding Living Systems. Cambridge University Press.

Editor's Summary: Dr. Denis Noble offered abundant evidence from 21st century biology showing that the reductionist gene-centric biology of the 20th century is largely misleading and incorrect. He emphasized that progress in contemporary medical science requires embracing an integrative organism-centered approach where complex lower-level functions are more clearly understood by considering their context in the higher-level functions of the organism. Dr. Noble also acknowledged that the cliches many people encounter in common experience like "we are born selfish," "it's in their DNA," and "they [genes] created us body and mind" — all suggested by Richard Dawkins — have negatively impacted society and influenced ideas in a wide array of other fields from economics to philosophy.

Since the turn of the century, when human DNA was first completely sequenced by the Human Genome Project, scientists have been looking for the "program" in DNA that creates the organism. But Dr. Noble explained that this is an illusion. As a computer programmer, he explained that programs must be in the form of "if-then-else clauses" to allow for choice and accommodate various possibilities. But there is no choice in the genome, no sequences of ACTG that are equivalent to "if-then-else clauses." DNA sequences have switches. Only when activated by the whole living cell are genes enabled to form a ribosome that will produce a protein necessary for a particular higher-level function. The central dogma of molecular biology claims that this process establishes natural upward causation from DNA to the whole organism.

Dr. Noble shared a recent article from the *British Medical Journal* showing that medical tests attempting to predict diseases like vascular or heart disease, Alzheimers, and cancer, by analyzing a few genes in the body, are largely ineffective. These polygenic risk scores produce as many false predictions as correct ones, thus he asserts that they are practically useless. The failure of these tests is due to genes being causally impotent. DNA is a chemical whose function is determined or constrained by higher-level living processes, so it cannot give reliable predictions of higher levels. DNA sequences cannot predict what constrains them, thus the conclusion of the article was not surprising.

He provided detailed examples of purposes coming from higher levels of the organism that influence constraints on lower levels such as (1) the role of the heart in the circulation of blood, (2) the role of kidney tubules in creating counter-current flow, (3) the role of Darwin's gemmules ensuring continuity of communication of characteristics in the organism and to the inheritance of later generations, and (4) Hodgkins cycles, i.e. the electrical activity of cells in an organism. Although genes are associated with certain bodily functions, it was emphasized that they are *not* causes of these higher-level functions. Based on the findings of a lifetime of experimentation, Dr. Noble argues for downward causation. This was depicted in a diagram of an inverted triangle — taken from his recent book *Understanding Living Systems* (2023) — where the highest level is the most open while the lowest is most constrained. The highest level was the sociotype ("our freedom to act, our decisions in relation to others in our societies," where "the mind is, in principle, the sociotype"), then the ecotype (niche), phenotype, organ types, tissue types, cell types, karyotype, and at the lowest level, DNA.

Dr. Noble made a point to give praise where it is due by acknowledging that understanding the double helix allowed scientists to learn about DNA replication. His presentation concluded with a sincere plea encouraging current and future generations of scientists and general society to seriously consider the negative influences of the reductionist gene-centric worldview and emphasized the necessity of progressing beyond it.

The following was not part of Dr. Noble's presentation, nor were the thoughts expressed herein addressed in the Q&A or inter-

disciplinary dialogue afterward. What follows was inspired by reflection on the conference after it finished.

A general principle underpinning Dr. Noble's approach, described in his book *Dance to the Tune of Life: Biological Relativity* (2017) as "answer[ing] questions at the level to which they are most appropriate and then us[ing] that insight to probe down and up towards the other levels," has significant implications for this conference's topic — considering life and cognition at the intersection of science, philosophy, and religion. In that book, he describes:

[O]ne purpose of the heart is to pump blood around the body to keep all the cells and tissues supplied with oxygen and energy [...] But it is equally clear that at the level of the atoms and molecules that form the heart, there seems to be no such purpose. Those atoms and molecules are the same as others in the body that do not form the heart. The purpose is not therefore intrinsic to the molecules. If one insists that all causation in biological systems is molecular, then there is no 'real' purpose anywhere in the body.

Just as questions about the maintenance of an organism's oxygen and energy supply must be asked at the level of a beating heart pumping blood throughout the body, where asking such questions and looking for answers at the level of atoms and molecules is ineffective, questions about the healthy maintenance of an organism's sleep quality must be asked at the level of daily life on Earth's surface. This is because circadian rhythm depends on a 24-hour cycle i.e. the amount of time between one sunrise and another observed from Earth's surface. In the exosphere (the outermost portion of Earth's atmosphere) where the International Space Station is situated, the time between one sunrise and another is 90 minutes, thus a 24-hour cycle is simulated with artificial light onboard the ISS for the sake of the astronauts' sleep quality. All questions regarding the experience of day and night, in general, must be asked at the level of being within a planet's atmosphere, as day and night simply do not exist in outer space. Beyond the solar system, even months and years cease to exist.

So, what level of organization is appropriate to answer questions like "Does God exist?" or "Is there a soul?" At the level of immediate sensuous experience (empirical observation), the answer seems to be no. That does not mean much, however, as seen in the previous two examples where answers to questions were being sought at an inappropriate level of organization. Even the existence of the heliocentric model of the solar system cannot be confirmed at this level. Direct observation sees the sun rising in the East and setting in the West, which suggests a geocentric model. Only through mediated thought and reason does one account for the spinning of Earth about its own axis as well as its orbit around the sun, and arrive at the heliocentric view. As will be discussed in Dr. B Mādhava Purī's presentation, answers to questions about the existence of the soul or God must be sought on a level of reasoning (organized/systematic thought) that accommodates the inherently paradoxical nature of reality as observed in basic phenomena like movement.

Superseding the Synaptic Network: How Cellular Complexity Transcends the Digital Neuron

Brian J Ford, Hon FRMS, Hon FLS | Cardiff University

Latest book: Nonscience Returns (2020)

About the Speaker: *Fellow of Cardiff University since 1986. Independent research biologist, author, and lecturer, publishing on scientific issues for the general public. Television personality for over 50 years. An international authority on the microscope. Was recently awarded the Ernst Abbe medal awarded by the New York Microscopical Society, honorary fellowship of the Royal Microscopical Society, and Fellowship honoris causa by the Linnean Society of London.*

Abstract: *Current digital models of cerebral function conceive of neurons as simplistic go- or no-go gates. They are regarded as organic* transistors. *My investigations over the decades have drawn a very different conclusion; namely, that the neuron is a largely autonomous cell of unimaginable complexity. Whereas current conceptions treat of neural networks, insisting that the centers of data processing lie with the synapses (and the large numbers of possible permutations which govern how they connect), my view is that data-processing is essentially intracellular. Rather than being discrete components in a grander network, neurons are envisioned as thinking for themselves. Examples of complex cognitive and sentient behavior drawn from the microbial realm substantiate the conclusion that cells are complex, and are far from the simple processing units we currently claim. It is not the brain that we can compare to a computer, the neurons being analogous to transistors. Rather, each individual neuron is far more sophisticated than the most advanced computer we know.*

Editor's Summary: Professor Brian J Ford began by vividly showcasing the sentient activities of cells through high-quality videos and photos. Some videos revealed ciliate microbes selectively choosing to consume certain items while rejecting others and Dino-flagellate algae unraveling an appendage called a flagellum to slowly and methodically regulate where it swims, while other recordings show white blood cells chasing bacteria, capturing them, and consuming them, such as the one taken in 1959 by David Rogers from Vanderbilt University. All of these prove that single cells are capable of intentional, purposeful, cognitive activity. Microscopists even observe single-celled algae like *Uroglena* possessing red eye spots that, when sectioned, show a curved retina. So, some cells can even see what they are doing. Other unicellular algae like *Spirogyra* and many diatoms engage in sexual reproduction utilizing mechanisms that are not understood. It is a mystery how mating diatoms sense and identify each other — where a protrusion emerges from each individual that touches the other and allows the two diatoms to fuse — and how they dissolve the cell wall which is necessary for such an act to occur.

Professor Ford criticized current textbooks for describing mating *Spirogyra* as simply putting out a conjugation tube, while the cognitive capacity of the algae to sense that they are near each other and determine which is male and which is female, and the mechanism behind how the conjugation tubes are formed and the cell wall is dissolved, are not addressed at all. He was also critical of the overly simplistic and inaccurate representations of cells, calling them "jelly bean models," that a large majority of academic institutions use to educate students and the public. Such models mislead people into thinking that cells are amenable to mechanistic descriptions, which reinforces the bias of modern science that everything is reducible to physics. If authentic images and videos of cells were utilized instead of jelly bean models, then the overwhelming complexity of living cells would be seen in truth and naive attempts of creating computer models of cells would be set in a proper context. Attempts to create digital models usually attract large sums of grant funding, however, so many scientists feel compelled to ignore real cells in favor of overly simplistic and inaccurate models.

Establishing that modern science lacks proper knowledge of relatively simple single-celled organisms served as a solid foundation for Professor Ford to show that science is completely at a loss for understanding neurons, or single nerve cells, which he postulates are "the most complicated and highly evolved cell in the universe." He critiqued modern neuroscience for focusing on neural synapses, where neurons meet but do not touch, instead of the nerve cell itself. Individual neurons do not just fire and send signals, they are self-regulated, autonomous, and cognitive entities in their own right. The synapses are basically just road junctions on a map; the focus should be on the people who live in the towns that the roads connect.

Pulling from a chapter that he contributed to *Brain, Mind, and Medicine: Neuroscience in the 18th Century* (2007), Professor Ford explained that in 1674, Antony van Leeuwenhoek was the first microbiologist to ever study nerve cells under the microscope. Nerves began to be depicted as mechanical devices, tubes filled with fluid, throughout the end of the 17th and beginning of the 18th century. Microscopist Johannes Evangelista Purkyně (1787-1869) was also a prominent individual who studied neurons under the microscope. The Purkinje cell was named after one of his detailed drawings of neurons, which were motivated by interest in dendrites coming out of the neuron beginning to lose focus while under the microscope. The drawings of Santiago Ramón y Cajal accurately depicted the complexity of dendrites branching out of a single neuron and even included major parts of the brain inclusive of each neuron and the pathways of its dendrites.

Professor Ford emphasized caution when confronted by claims of modern neuroscience, as there are clear cases where knowledge claims were embellished to ensure funding of large-scale projects. Examples include (1) Artechouse's "Life of a Neuron" exhibit, produced in partnership with the Society for Neuroscience, which claimed to depict the entire life of a neuron and reveal mysteries of the human brain, although the vivid artistic computer renderings did not factually deliver on that promise, and (2) the Human Brain Project, which cost €1-billion funded by the European Union, producing many publications and professional gatherings that all suffer the defect of telling very little about the way living neurons actually function. The presentation concluded by advocating humility among scientists, suggesting that academic courses dedicate Fridays to considering what is not known about particular subjects, rather than what we do, to inspire humility in teachers and creativity in students.

3

Death & Desire: Negativity at the Foundation of Life

B Mādhava Purī, PhD | Princeton Bhakti Vedanta Institute

Latest book: Idols of the Mind vs. True Reality (2020)

About the Speaker: *Received PhD in Theoretical Chemistry from Georgetown University. Postdoc at the National Bureau of Standards in Washington DC. Published technical papers in The Journal of Chemical Physics. Turned to the Indian school of yoga to learn about consciousness. Started GWFHegel.org. Serving Director of the Princeton Bhakti Vedanta Institute. Visionary behind this annual conference series since 2013.*

Abstract: *At the root of all life is negativity, a self (concept, identity) which is the same as the negation of what is other than itself, it's object. This negation is not annihilation of the other but as life it is a unity or identity of self-concept and its object. This is how Hegel presents the logical structure of life. Positivist approaches understand* *living wholes through the lens of mechanism as mere aggregates of externally assembled parts.* Approaches embracing negativity recognize the mutually dependent bi-directional internal causal relation among various parts and between parts and whole. *A living individual is not just a unitary positive being but simultaneously a negative or differentiated multiplicity, thus the production and dissolution of itself facilitates growth and development as a coherent whole.* The boundary of the individual's being for itself and what is other than itself gives rise to a relation as a living process. Determining what something is, also entails what it is not, i.e. taller implies not shorter and day implies not night. Hegel explains that the negative subjective aspect of the objective content of living entities is the concept, i.e. that form which determines/contains an objective content. These two aspects mutually interpenetrate each other in dynamic dialectic relation as an identity-in-difference. Due to the discrepancy between the living individual's being-for-itself and general being-in-itself, it experiences an internal lack, i.e. desire to overcome the discrepancy through reciprocal activity with the environment. One such need or necessity that arises is the unity of its individual life with the same life principle of other organisms – that manifests in nature as reproduction, which perpetuates as the universal genus process. Death of a particular organism emphasizes the universal or collective essence that both transcends and is immanent in all living individuals, which in the externality of nature manifests as the idea of soul in a body.

Editor's Summary: Dr. B Mādhava Purī introduced a logical understanding of life beyond the mere molecular mechanistic ideology embraced by modern science for the last 400 years. Since the removal of Aristotle's formal and final aspects of cause from modern science, the role of the scientist's rational thinking in formulating empirical scientific knowledge is usually not explicitly addressed in explanations of science. This is the underlying motivation for the Science & Scientist conference series. Utilizing G.W.F. Hegel's philosophy offers a way to connect rational activity with natural phenomena.

Just as mathematics provides insights about material things beyond empirical observation and positivist thinking, similarly, dialectic reasoning — which accommodates paradox/contradiction as fundamental to reality — offers deeper insights into the role of negativity as it pertains to living cognitive phenomena. The negation of the physical (objective constitutive content) is not nothing — it is the non-physical (subjective concept). Life is an existing contradiction because it is the negation of itself; it is what it is not. Life is essentially the continuous activity of seeking what it lacks and satisfying what it needs. This dynamic organic becoming transcends reified notions of static being. Dr. Purī justified that paradoxes exist in nature by showing that basic movement is itself contradictory. To move through a point in space means that the moving object simultaneously is and is not in that location. The dynamic activity of passing through a point must be conceived as a continuous process rather than being reduced to a static moment of that process. Only then can the paradoxical nature of movement be genuinely comprehended.

The relationship between an organism and the environment is also paradoxical. They form an *identity-in-difference*, due to their *identity* in being intrinsically interdependent where organisms derive sustenance from the environment while the environment's very existence is comprised of diverse organisms (plants, animals, fungi, etc) and elemental conditions (terrain, climate, sunshine, etc), which also implies *difference* between them. To the extent that an organism feels incomplete within itself, it experiences the desire to exploit the environment to fulfill its inner lacking. This manifests as hunger, thirst, sexual arousal, and ambition, which are all responses to the negative activity of needing something other than or external to oneself. This can be positively described as purpose, where the fulfillment of particular desires becomes the reason that an organism acts in a certain way towards the environment.

Hegel describes three levels in the Logical development of Life, where life encompasses them all. These emphasize life's inherently conceptual nature.

1. The living individual — the organism as such
2. The living process — the organism's relation to the environment
3. The genus process — the organism's relation to other organisms in the reproductive act, where death plays the important role of demonstrating the fleeting nature of the particular organism under the universal genus, which is intrinsic to and inseparable from the organism

Science affirms that, at the cellular level, we change bodies several times throughout a single lifetime, yet we remain the same person. We remember circumstances from our entire life whether something happened while inhabiting a youthful body or an aged one. So, the unchanging aspect of living entities is the concept or the self/soul, and the aspect that constantly changes is the constituent content or the body. The soul is the negation of the body, but negation does not mean annihilation. The negation of 1 is not 0, it is -1. The negation of the physical is not nothing, it is the non-physical. This non-physical unchanging soul unifies/integrates the diverse constantly changing physical bodily constituents (one person), as well as various bodily activities throughout time (one person's memory).

In nature, the self constantly works to satisfy bodily needs that force its attention outwards toward the environment in search of food, shelter, and mates. Bodily neglect causes unpleasant inner experiences, thus the self only satisfies bodily needs to feel fulfilled within itself. But inner fulfillment based on bodily satisfaction is temporary. We eat and get hungry later. We sleep but eventually get tired again. Even the vigor of good health eventually subsides. The body cannot maintain the standard of satisfaction that the self desires. Religion and philosophy offer humankind systematic knowledge about how to permanently achieve inner fulfillment and what happens to the soul after the body perishes.

In the Logical development of the Absolute Idea, Hegel shows that there is a dialectical stage in which the concept is immediately identified with its constituent content. This is called Life. In nature, this corresponds to an identification of concept/soul and body. The world is then cognized through this incomplete notion, and as a result, cognition is solely attributed to the body, i.e. scientists think that consciousness is a product of a physical event like firing neurons or vibrating microtubules. Overcoming this kind of positivist thinking requires embracing the role of dialectic negativity. The relationship between self/soul and body is not a mere positive identity, but an identity-in-difference, where the self is the negation of the body and vice versa. This negation produces something real, but the reality of what is produced must be known at the appropriate level of organization. The physical body (constitutive content) can be known at the level of sensuous experience, while the non-physical soul (concept) is known at a higher level of pure thought and reason. The idea that lower levels of organization are constrained (determined) by higher more open levels, as Dr. Denis Noble mentioned, is maintained here.

Despite what molecular mechanistic ideology would have us believe, living wholes are not an aggregate of externally assembled parts. The whole and parts have a reciprocal causal relation motivated by internal purpose. This is observed in the ontogeny of organisms and was recognized by Immanuel Kant in his *Critique of Judgement* (1790). Thus, scientists cannot just inquire about parts. The whole of which a part is constitutive — that for the sake of which it exists — must be considered for a sober understanding of any given part. Currently, most scientists do not recognize this, so they think that a carbon atom at the tip of their nose is the same as the one in the leg of an elephant. Such atomic monism is an abstract understanding, thus it is inappropriate for a comprehensive knowledge of living entities due to their intrinsically dynamic organic nature.

Dr. Purī concluded by expressing that his presentation was meant to suggest a general direction, encouraging people to study Hegel and seriously consider the relevance of his ideas to contemporary biology. He candidly acknowledged the difficulty in explaining molecular phenomena in terms of Hegelian philosophy, since Hegel was not aware of molecular biology, although a hint to overcoming this problem may be found in his anthropology. When the intrinsic subjectivity of life and the capacity for self-consciousness of the human being are properly acknowledged, then the need to go beyond mechanical, chemical, and even teleological explanations becomes apparent. While teleological explanations get us beyond reductionist thinking by encouraging systems thinking, it is necessary to acknowledge that systems are not self-conscious. This is evident when contemporary scientists like synthetic biologist Michael Levin utilize systems thinking to define "self" within a teleological (goal-directed) framework concerned with "functional, third-person, objective capacities, computations, and behaviors," while intentionally avoiding "first-person experience or a sense of self as qualia," i.e. a first-person account of selfhood (self-consciousness). Progress here requires engaging with dialectical reasoning that accounts for negativity at the foundation of life. This is the level of organization at which answers to questions about the existence of the soul or God must be sought.

Understanding the Vedāntic View on the Difference between Life & Non-Life

B Niskām Śhānta, PhD | Sri Chaitanya Saraswat Institute

About the Speaker: *Received PhD in Coastal Hydrodynamics from the Indian Institute of Technology - Kharagpur. Postdoc at the Korea Ocean Research and Development Center. Published numerous papers in international/national conferences and journals like Springer Link and Communicative & Integrative Biology (PMC). Main organizer of this Science & Scientist conference series since 2013. Vaiṣṇava monk. Sevaite-President-Acharya of Sri Chaitanya Saraswat Math in Narasimhapalli (Nabadwip Dham), West Bengal, India.*

Abstract: *The conceptual differences between life and non-life are very difficult to understand within the framework of modern materialistic science. Modern science has painted an image of life as a very complex molecular arrangement, and thus there is no major difference* *between life and non-life except for an increase in complexity in molecular arrangements in the case of life. Despite that, science could never demonstrate how life can appear from non-life by increasing the complexity of molecular arrangements. However, in the Bhāgavat Vedāntic tradition, the knowledge that we have of reality is very much dependent on our attitude. The concept of matter, or non-life, is an experience of souls who are not fully surrendered to the divine Absolute, and such souls live a material life in this material world. There is another reality that is transcendental to this material world, and there is no such duality that we experience as life and non-life. In this talk, the speaker wants to highlight these ideas with further details.*

Editor's Summary: Dr. B Niṣkām Śhānta began by acknowledging that modern science tries to define life in terms of nonliving things. He briefly discussed the history behind the idea of abiogenesis (the spontaneous generation of life from insentient matter) explaining that it was disproved by Louis Pasteur's swan-neck flask experiment, and highlighted the inconsistencies of how Stanley Miller's intelligently executed experiments constructing molecular building blocks of life could support an abiogenic origin of life hypothesis. Miller once conceded that even if he was given all of the organic compounds necessary for life, he was unsure if a living cell could be produced from assembling them.

Recognizing that reductionist ideology fails to address questions like the minimum number of parts essential for the survival of an organism and the mechanism responsible for assembling these parts, Dr. Śhānta criticized the notion of Darwinian evolution of insentient bodies. He (1) showed self-defeating doubts held by Charles Darwin and Francis Crick contemplating if a mind evolved from a monkey could be trusted to produce a true theory of reality i.e. Darwinian evolution, (2) pointed out discrepancies between the fossil-based and genetics-based tree of life, and (3) challenged the central dogma of molecular biology (unidirectional causality from DNA → RNA → Protein [Enzyme] → Trait) by referring to whole-part and part-part reciprocal circular causality.

As Ernst Mayr stated, the mechanistic view of an organism is exceedingly superficial. Living entities act as self-determined independent conscious agents, as acknowledged by the Cambridge Declaration of Consciousness — a product of the Francis Crick Memorial Conference in 2012 — which recognized the capacity for intentional behavior of non-human animals based on conclusions of neuroscience. This freedom is what distinguishes living entities from material objects including AI machines. In this regard, Dr. Shānta went on to consider different perspectives on the origin of consciousness and the nature of the self.

The Vedāntic view suggests that the purpose of life is to inquire about the self and that all education and research should be dovetailed in this direction. The conclusion of empirical science — which emphasizes that objects are independent of subjects — is that consciousness is a product of matter. Vedic wisdom recognizes that the existence of a conscious observer logically precedes all sensuous experience and empirical knowledge, and suggests epistemological approaches beyond empiricism that are more reliable than faulty human senses. *Bhagavad-gītā* verse 13.34 describes consciousness as the empirical symptom of the soul, where consciousness emanates from the soul and illuminates its worldly experience just as sunshine emanates from the sun and illuminates the world. The role of the subject — the conscious living entity — and how its state of consciousness determines its perception of objects, is emphasized. Here, subject and object have an interpenetrating dependent relationship such that our particular level of conscious development influences what plane of reality our activities are being conducted on, i.e. determines what world we live in.

Vedic knowledge offers detailed descriptions of different categories of conscious development that living entities experience throughout various lifeforms: covered consciousness like trees (*ācchādita-cetana*), shrunken consciousness like non-human animals (*saṅkucita-cetana*), primitive human consciousness like aboriginals (*mukulita-cetana*), general human consciousness where religion, philosophy, science, art, and government emerge (*vikasita-cetana*), and the consciousness of a self-realized saintly person surrendered to the Absolute Truth (*pūrṇa-vikasita-cetana*). In this context, Dr. Śhānta emphasized the difference between general religious practitioners and those rare individuals who have factually developed a deep realization of the Truth.

The Vedic schools of *aṣṭāṅga-yoga* and Advaita Vedānta discuss four states of consciousness that a person can experience. Herein, the subtle body refers to that which is constituted by intellect, false ego, and mind, while the gross body is constituted by the five elements — ether, air, fire, water, and earth. There is the waking state where both the gross and subtle bodies are active (*jāgrat*), the state of dreaming sleep where the gross body is inactive and the subtle body is active (*svapna*), dreamless deep sleep where both the gross and subtle bodies are inactive (*suṣupti*), and pure consciousness or the state of liberation which transcends both gross and subtle bodies (*turīya*). The first three states are considered the plane of exploitation, where people's activities are primarily motivated by self-centered interest, and the fourth state is the plane of renunciation motivated by exhaustion from suffering the reactions of exploitive activity over many lifetimes. The Bhāgavat Vedānta philosophy of the Vaiṣṇavas mentions a fifth state (*turīyātīta*), where the individual does not just passively occupy the liberated state, but learns how to actively engage in transcendental spiritual activities — the plane of dedication. Within the plane of dedication, there are further variegated planes of reality depending on the particular mood of heartfelt devotion that an individual embodies and is attracted to.

In this way, ancient Vedic wisdom offers systematic knowledge explaining the conscious development of the non-physical soul through differentiated forms that are determined by an individual's sincere desire. The degree to which a particular bodily form accommodates the full expression of the soul's conscious capacity — whether in the body of an aquatic, plant, insect, reptile, bird, animal, or human — is determined by the soul's specific desires and the degree to which these correspond to the interest of the Absolute Truth. Rather than the evolution of insentient bodies proposed by Darwin, the Vedāntic view vividly describes an evolution of consciousness.

From this point of view, the trouble that modern science faces in distinguishing life from non-life (matter) is a result of polluted consciousness. The variety found in the material world is described as the perverted reflection of the higher spiritual world, which is also variegated. The main difference is that spiritual variety is properly situated by being centered on the Absolute Truth — the Organic Whole, the Supreme Personality of Godhead, Bhagavān Śrī Kṛṣṇa — Who is known by many names. Since Kṛṣṇa, as the Absolute, is intrinsically all-accommodating, all-pervasive, and all-attractive, the spiritual world is harmonious and all individuals are situated in their constitutional position. On the other hand, in the material world, each living entity has forgotten their eternal relationship as a loving servant of God and instead has established themself as the center of their particular experienced reality. The discrepancy between this materialistic worldview and the Truth produces an illusion, a distorted perception of reality, where humans fight amongst themselves to establish their specific self-centered point of view to secure temporary comfort and pleasure. This material suffering is equivalent to the experience of a fish out of water — we are not meant to live in the plane of exploitation. Our natural environment as souls is the spiritual world, the plane of dedication, where all activities are dovetailed with the higher purpose of satisfying the Supreme Lord.

Dr. Shānta concluded by explaining that attitude and knowledge are related. A self-centered attitude dependent on our defective senses and limited mind produces relative knowledge that seems true for some time but eventually becomes outdated and is replaced. He explained that even knowledge gained through yogic meditation is limited by our inherently finite nature, and encouraged further study of Vedic epistemology. Vedic epistemology describes methods like direct sense apprehension (*pratyakṣa*), learning from someone else's direct sense apprehension (*parokṣa*), mystic knowledge gained through meditation (*aparokṣa*), transcendental knowledge that is completely different from mundane experience in the material world (*adhokṣaja*), and transcendental knowledge that appears similar to experience in the material world although it is not mundane (*aprākṛta*). Founder-Āchārya of the Śrī Chaitanya Sāraswat Maṭh, Śrīla Bhakti Rakṣak Śrīdhar Mahārāj, explains the necessity of *aprākṛta*:

That is Goloka, the full-fledged theistic conception which is only found in Kṛṣṇa's domain. Central knowledge of the Absolute must have a connection with even the lowest level of mundane creation; it must be able to harmonize the worst portion of the illusory world. This is known as *aprākṛta*, supramundane. To enter that highest realm is possible only through divine love.

The first three kinds of Vedic epistemology — *pratyakṣa*, *parokṣa*, and *aparokṣa* — are ascending approaches where the finite individual tries to forcefully attain knowledge through asserting its infinitesimal power. The last two kinds — *adhokṣaja* and *aprākṛta* — are descending approaches (*avaroha*). Complete or perfect knowledge of the Truth must be revealed, descending from the infinite to the finite, which requires surrender by the finite entity. This means that the finite individual must abnegate their false ego and all attempts to act as an independent entity. The practical process for doing this, gradually purifying our intelligence and cultivating transcendental spiritual vision, is called *bhakti-yoga*, where emphasis is placed on proper association. Being around certain people can condition us towards a lower state of consciousness while associating with others promotes higher-level thinking and consciousness. Thus, the practice of *bhakti-yoga* is carried out under the affectionate guidance of the bonafide self-realized preceptor (Guru) who comes from a succession or lineage of other bonafide teachers (Guru-paramparā). The degree to which a sincere seeker makes progress depends on the extent to which they thoughtfully surrender, devoting time and energy to humbly inquiring from and rendering service to the Guru.

5

Emotional Intelligence in Cows

B Vijñān Muni, PhD | Sri Chaitanya Saraswat Institute

About the Speaker: *Received PhD in Chemical Engineering from the Indian Institute of Technology - Kharagpur. Published peer-reviewed papers and book chapters in international conferences and journals like Springer Link. President of the Sri Chaitanya Saraswat Institute based in West Bengal, India. Vaiṣṇava monk.*

Abstract: *In the Vedāntic view of ancient Indic traditions, animals like cows, elephants, horses, etc., have a very high position in terms of their closeness to human beings and developmental stage among biological forms through the cycle of reincarnation. Cows give auspicious products like milk, cow dung, urine, curd, and Gorochana – all of which give benefits to human beings. In the Vedic view when cows are taken care, it reduces quarrels and sinful propensities among human beings. Ancient Indic societies maintained millions of cows in every village and state. Modern scientific and behavioural studies are also establishing that cows have well developed cognitive and learning* *abilities. Scientists have recorded changes in vocalizations and pulses according to different happy or stressful emotional conditions such as cognitive bias, emotional contagion, and mother-calf bonds. Cows are social animals and even human beings experience emotional benefits from them. In the Vedas it is mentioned that cows are one of the universal mothers. Many Ayurvedic medicines are prepared from cow milk. The emotional nature of cows can be compared with those of other domestic animals such as buffaloes, dogs, goats, etc., to understand their unique and superior qualities. The Vedic view of living forms is that of organic whole and unity, yet cows have a higher social and spiritual position. Cows should be protected by governments and should never be killed. With proper care human society can gain all kinds of economic prosperity with more realizations of the potential of cow products in medicine, food, agriculture, and emotional well-being among humans and the environment.*

Editor's Summary: Dr. B Vijñān Muni began by explaining that he has had the opportunity to intimately observe the behavior of cows in the cowshed (*gośālā*) at the Śrī Chaitanya Sāraswat Maṭh temple located in Narasimha Palli, West Bengal, India. He shared a personal experience of being with one cow mother who produced a lot of milk. Despite the fact that most of her milk was utilized for the temple's purposes and only a small portion was given to her calf, this cow mother interacted with Dr. Muni in a loving, affectionate, and non-envious manner.

Dr. Muni reiterated the acceptance of non-human animals as conscious entities by the Cambridge Declaration of Consciousness and the descriptions of various Vedic categories of conscious development observed throughout lifeforms like *ācchādita-cetana*, *saṅkucita-cetana, mukulita-cetana, vikasita-cetana,* and *pūrṇa-vikasita-cetana.*

India's traditional Vedic culture heavily emphasizes the importance of cows as central to human development and prosperity. Many preparations are made from cow milk that continue to be used in Vedic rituals and sacrifices. The ancient *Yajur-veda* explains that caring for cows increases the duration of a person's life while the *Atharva-veda* says that as long as the sun shines in the universe, cows will exist and that the whole universe depends on the support of the cow. In this regard, Dr. Muni acknowledged that even today, civilizations throughout the world — whether or not they are vegetarian — depend on cow dairy products.

Presently, due to their being heavily exploited for meat production, the question of how intellectually and emotionally sensitive cows are has become pertinent. Modern scientific research confirms that cows possess learning capabilities and acute sensory perception such that their auditory capacity is superior to horses, olfactory capacity plays an important role in social life, response to tactile sensation significantly influences their emotional state, and sophisticated visual discrimination allows them to distinguish familiar things. Cows display an advanced capacity for recognition where mature cows distinguish between familiar handlers and unfamiliar handlers despite the same uniform being worn. They even discriminate between photographs of cows and other species of animals, where the sameness among varying kinds of cows was identified despite the phenotypic variability.

It is accepted that many non-human mammals experience basic emotions — understood as subjective, behavioral, neurophysiological, and cognitive processes that shape attention, decision-making, and memory. In non-mammals, the term "affective state" seems to be used in place of "emotion" to signify a less sophisticated capacity. This suggests that emotions play an important role in the behavior of cows including being a foundation for more complex capacities like being empathetic and self-aware. Cows display particular behaviors when they are afraid, such as being apprehensive to pass stool, using specific vocalizations, and trying to run away. If they become stressed from a lack of food, the percentage of white in their eye decreases. Nasal temperature, ear posture, and heart rate are also influenced by the emotional state of a cow. When they are happy, cows become playful.

Complex emotional reactions suggest that cows have advanced capacities for self-awareness and empathy. (1) Emotional reactions to learning — cows' reactions after performing an activity better suggested that they recognized their improvement and became emotionally aroused by this, suggesting self-awareness. (2) Cognitive bias — after a negative experience like being branded by a hot iron, young Holstein calves were less likely to approach ambiguous objects in their environment for up to 22 hours, suggesting caution inspired by trauma. (3) Emotional contagion — cows are often observed sharing their emotions, where just by observing one cow becoming aroused by a specific emotion another cow will exhibit the same. (4) Social buffering — cows are highly dependent on the companionship of family/herd members and can more easily deal with environmental stress when they are together. One instance of social buffering is calves being regularly licked by their mother becoming healthier than those who are not — a phenomenon observed directly by Dr. Muni at the SCS Maṭh cowshed. He concluded by explaining that the progress of 21st-century biology has revealed that organisms are not replaceable mechanical units in a set of species members, but that they are unique individuals that must be increasingly understood in terms of selfhood and personality.

The Problem of Embodied Consciousness in the Lens of Vedāntic View of Consciousness

Rajakishore Nath, PhD | Indian Institute of Technology (IIT) - Bombay

About the Speaker: *Professor of Philosophy at IIT Bombay specializing in Artificial Intelligence, Philosophy of Mind and Cognitive Science. Published many peerreviewed papers. On the Advisory Board for AI & Society: Knowledge, Culture and Communication by Springer.*

Abstract: *I would like to discuss embodied consciousness in the lens of Vedāntic perspective. The Vedāntic perspectives of consciousness, especially Advaita and Viśiṣṭādvaita. Advaita perspective counters the embodied consciousness. The embodied theory consciousness has offered causal explanation to consciousness. The causal explanations are very* *much integration with the state of the organism with the environment. That is to say that the human consciousness depends on bodily gestures around the world. This kind explanation is based on the ground that consciousness is causally dependent on the material universe and that all conscious phenomena can be explained by mapping the physical universe. In this regard, consciousness is basically a bodily phenomenon and can be mechanically explained following the naturalistic methods of science. On the other hand, the Advaitins are proposing the concepts of Brahman and Ātman that are Absolute and are the source of all form of consciousness and the world. The pure consciousness, i.e., Ātman, cannot be 'known as an object of mediate knowledge, yet it is known as involved in every act of knowing'. Therefore, the nature of consciousness is transcendental. This transcendental nature of consciousness is ontologically real and very difficult to explain in the theory of embodied consciousness.*

Editor's Summary: Dr. Rajakishore Nath understood theories of embodied cognition/consciousness to refer to reductionist explanations that do not give consciousness any ontological status and refer to it as an epiphenomenon of matter. He explained that the inspiration for the idea of embodied cognition traces its origins to Descartes' disembodied theory of mind, where mind and body were seen as two distinct and independent substances. In what seems to be a refutation of Cartesian dualism, embodied cognition claims that consciousness is inherently connected to its surrounding environment and is causally dependent on the material universe, thus it is amenable to naturalistic descriptions by modern science. Dr. Nath was unsatisfied with the reductionist foundation of the naturalistic descriptions of consciousness advocated by contemporary philosophers like Daniel Dennet, John Searle, and David Chalmers.

He criticized the notion of "access consciousness," which emphasizes a third-person perspective of conscious experience, as being abstract and lacking the concreteness of first-person phenomenological accounts of consciousness, and argued further that it is more appropriate to consider consciousness as an irreducible subjective qualitative field of experience.

Dr. Nath distinguished between bodily or embodied consciousness, which he referred to as micro and dependent on certain circumstances, and transcendental pure consciousness, referred to as macro and independent of worldly causality. He takes this to mean that universal pure consciousness is self-manifesting and that it must be independent of all objects, emphasizing this contentless aspect that is unintentional and not the "consciousness of" any object. Dr. Nath advocated for serious consideration of the Advaita Vedānta view, especially that of Śrī Aurobindo. He rejected the Viśiṣṭa-Advaita view of consciousness being an active, intentional, dynamic quality of the self and said that the idea of a relationship between consciousness and the self is dismissed by Advaitins like Aurobindo due to its contradictory nature. These Advaitins hold that a substance (Brahman) must be either identical with or different from its attributes. They believe that consciousness is the only reality and that consciousness of the world is an illusion experienced when the former is not yet realized, while Viśiṣṭa-Advaita posits that there is genuine difference as well as identity between consciousness and the world.

Dr. Nath concluded by emphasizing that both naturalistic explanations reducing consciousness to matter and approaches holding the world and consciousness in juxtaposition are not sufficient. Since Advaita rejects any kind of reduction of consciousness to material objects, he advocated the approach of Advaitins like Aurobindo who collapse the aforementioned juxtaposition to an undifferentiated identity.

What follows was not part of Dr. Nath's presentation, but is an elaboration of relevant points addressed by Dr. B. V. Muni during this session's Q&A and elaborated in Dr. B. M. Puri's talk.

Most modern scientists and philosophers understand contradiction/paradox to be a symptom of falsity. If two phenomena are described as being the reciprocal cause of each other, like protein and DNA, or if one claims that substance is both identical with and different from its attributes, many thinkers dismiss such claims based on their paradoxical nature. Such people resort to conclusions based on either dualistic or monistic/non-dualistic thinking. Dualistic metaphysics holds mind and matter or thought and being in a juxtaposition where they are causally and existentially independent of each other. Monistic/non-dualistic metaphysics either reduces mind/thought to matter/being (materialism) or matter/being to mind/thought (idealism), and can lead to voidism. Conclusions being formed by resorting to either dual or monistic metaphysics has given rise to the entire gamut of philosophical and scientific theories throughout Eastern and Western history. Higher-developed dialectical reasoning — which accommodates contradiction as a necessary moment of dynamic development — can help 21st-century science and philosophy avoid this historical pitfall.

Journey into Mind - DNA - Consciousness

Anandi Ravinath, MS | Inner Light Foundation, Mumbai, India

Latest book: MIND Your DNA (2019)

About the Speaker: *Received MS in Biotechnology from IIT Bombay. Trustee with Inner Light Foundation, Mumbai, India. Worked for pharmaceutical and biotechnology companies in Germany and India.*

Abstract: *Scientists and Spiritual seekers are searching for the "Truth." Scientists uses gadgets to probe deep into the physical matrix of the manifest universe, while Spiritual seekers search for answers from within oneself and use the Human Mind & Body as the laboratory. It is important to appreciate the Science in Spirituality because spiritual growth or Self-Liberation is a scientific process. Despite scientific advancement, the intricate workings of Human brain, mind & consciousness is still a mystery. Epigenetic changes have been attributed* to various diseases in the recent past and reversal of these epigenetic changes have also been reported by various scientific studies through relaxation techniques like Meditation. Thus, there is a scientific proof that Stress impacts the DNA negatively, while Meditation impacts the DNA positively. It is also a well-known fact that stress is felt by the MIND before the physical body feels its impact; same goes with Meditation where the relaxation and clarity is first felt by the MIND before positive changes are noticed in the physical wellbeing. This is a clear scientific proof that MIND indeed is in our DNA. Mind is closely associated with our Consciousness and so needless to say that DNA, MIND and CONSCIOUSNESS are in constant communion.

Editor's Summary: Anandi Ravinath presented ideas about the relationship between mind, DNA, and consciousness. After witnessing people who attended Kumar Krishnamoorthy's meditative programs experience relief from physical ailments, Ms. Ravinath hypothesized that meditation affects the body at the molecular level. Through continued meditation on this topic (the Vedic epistemological method of *aparokṣa*), further intuitive insights were gained which she tried to substantiate through Vedic texts and existing scientific evidence.

Ms. Ravinath distinguished universal all-accommodating consciousness from mind, an individual network of thoughts like memories, where activities of the mind can influence gene expression. She justified an intimate connection between mind and DNA by explaining that mind is the energetic component that corresponds to the physical matter of DNA. The three-dimensional structure of DNA accommodates information storage, just as information is stored on paper through written language or in digital devices through code.

Research at Harvard University's Wyss Institute has proven DNA's tremendous information storage capacity. Ms. Ravinath explained that DNA not only contains vital information for the development of organisms but also stores new heritable information (through epigenetic inheritance) throughout the unexpected circumstances of an organism's lifetime.

The top-down relationship between mind and DNA was substantiated by citing research explaining that emotional responses to the environment, like stress, directly impact people's genome and can induce long-term epigenetic changes like mental disorders. Conversely, meditation has been shown to improve immune response and decrease levels of stress, thus positively impacting the genome. The work of scientists like Alexander Gurwitsch (1874-1954), Dr. Fritz Albert Popp (1938-2018), and Dr. Roeland Van Wijk shows that biophotons possibly originating from DNA are emitted from cells. These biophotons help with intracellular communication. For Ms. Ravinath, this photon light emission is the presence of consciousness in DNA. Dr. Roeland Van Wijk's research shows that apparently, meditation decreases biophoton emissions. For Ms. Ravinath, the correlation between the cessation of thoughts during meditation and the decrease of biophotons validates her hypothesis about the union between biophotons and consciousness.

She concluded by encouraging future research efforts to embrace introspective meditation as an approach to gaining knowledge, in addition to research by external (third-person) observation done in laboratories.

What follows is an elaboration of points briefly addressed during the conference by Krishna Keshava Das in this session's Q&A, in addition to elaborating on a point addressed by Dr. B Mādhava Purī while answering a question posed by Ms. Ravinath during his talk.

Distinguishing between life and non-life (matter) is integral to further considering the relationship between matter, life, and mind/consciousness/cognition. While cells and organisms are alive, DNA is not. It is a lifeless chemical whose function is determined or constrained by higher-level living processes in cells and organisms. Nobel laureate Barbara McClintock described the genome (DNA) as "a highly sensitive organ of the cell." Hype following the Human Genome Project inspired scientists to seek a "program" in DNA that creates the organism. Dr. Denis Noble explained that this will never happen because programs must be expressed in a way that allows for choice and accommodates various possibilities, but there is no choice in the genome. Sequences of ACTG are not equivalent to meaningful expressions. They do have switches, however, which only function when activated by the whole living cell. DNA molecules are not autonomous agents, rather it is living cells and organisms that possess cognitive capabilities and exert agency over the genome.

Notions that "thoughts are photonic in nature" or that biophotons are consciousness are just as influenced by positivistic understanding as notions that consciousness is produced from firing neurons or vibrating microtubules. Such ideas immediately identify matter and consciousness without conceiving their simultaneous difference. As Dr. B Mādhava Purī suggested, a deeper more mediated comprehension of the relationship between matter (physical) and consciousness (non-physical) requires considering the role of negativity through dialectic reason.

Utilizing both epistemological approaches of external empirical observation and internal contemplative meditation requires establishing a standard of accountability to ensure the integration of accurate systematic knowledge across approaches. Once data from the natural world has been carefully gathered through empirical observation, meditation can be utilized to refine comprehensive conclusions. Introspective meditation helps cultivate self-consciousness — a clear sense of the necessary contribution that the self inherently makes to knowledge, i.e. the role of the scientist in the development of scientific knowledge. A self-conscious scientist may consider the following questions:

- What axioms are employed in scientific experimentation?
- What presumptions shape the boundary conditions for particular experiments?
- What biases influence discrimination between noise and signal in data collection?
- Is there a necessity for such axioms, presumptions, and biases?

Recognition of the intrinsic role of the self in the production of empiric knowledge may serve to disillusion scientists who believe that they occupy a third-person perspective as passive observers of nature. The thoughts of an observer are intimately entangled with what is being observed and comprehensive knowledge should soberly reflect this. Formulating a scientific method that accommodates a first-person phenomenological account of how purely logical thought determinations serve as a basis for further determined sensuous thoughts about natural phenomena would integrate empirical (external observation) and meditative (internal observation) approaches. This would widen the epistemological scope through which modern science can discover the truth.

Evolution is Cognitive Thermodynamics

J Scott Turner, PhD | State University of New York (SUNY)

Latest book: *Purpose and Desire (2017)*

About the Speaker: *Emeritus Professor of Biology at SUNY. Project Director on science-related issues for the National Association of Scholars. Physiologist by training, but with a deep interest in the interface of physiology with evolution, ecology and adaptation. Prolific author of peer-reviewed scientific articles and books.*

Abstract: *Charles Darwin sought a natural law explanation for the evolution of life, which he hoped would be free of the vitalist predilections of his grandfather, Erasmus Darwin, and his French predecessor, Jean-Baptiste Lamarck. His solution was natural selection: variation of form and function within generations, with some variants being "naturally selected" for success in breeding. Darwin's conception of natural selection was inextricably bound up with the* *organism, particularly in the phenomenon of adaptation, for which he constructed an elaborate theory of heritable adaptation, pangenesis. In the 1920s, Darwin's organism-centered concept of adaptation was replaced by a gene-centered concept, which conferred fitness on genes for "apt" function relative to genes for "inapt" function. As a result, the vital phenomenon of adaptation was lost.*

A coherent theory of evolution requires a coherent theory of life, which modern Darwinism lacks. What is needed is an explicit recognition of life's unique attributes, among them cognition, intentionality, purposefulness, and creativity, but grounded in life as a thermodynamic phenomenon. I argue that recognizing life's unique properties ("small-v vitalism") is not only compatible with understanding life as a thermodynamic phenomenon, it provides a more coherent theory of adaptation and evolutionary change. This negates the Darwinian conception of evolution, however, because it makes evolution a profoundly purposeful phenomenon, driven by the intentionality and creativity of life.

Editor's Summary: Dr. J Scott Turner is interested in what physiology (the science of how life works) has to say about how life evolves. He discussed a series of eight conundrums that emerged from lifelong research on this topic. A photo of a large mound built by a colony of termites was the first conundrum. The mound — which can exceed 9 feet (≈3 meters) in height and is composed of around 750 pounds (≈340 kilograms) of soil carried by the colony members — serves as an above-ground wind-driven lung for a colony living deep underground. The form of the mound corresponds to a particular function, and the form changes in various circumstances. Thus, Dr. Turner identifies living attributes in the mound like (1) functionality, (2) adaptability, and (3) design. The termites, colony (superorganism), and mound serve as an integrated "extended organism" which "[act] as a coherent living system." He emphasized that the environment is an active participant in the life of an organism, not just simply a place to live.

The second conundrum regarded the contemporary "cybernetic interpretation" of homeostasis versus what Claude Bernard, the founder of modern physiology, actually said. The cybernetic interpretation says that regulatory "machines" maintain an organism's internal environment like temperature, salt concentrations, and metabolic processes, such that homeostasis is an outcome of this mechanical activity. But Bernard said that "The steadiness of the internal environment is the condition for a free and independent life" such that homeostasis is fundamental to the living process. He held that homeostasis is unique to life and distinguishes it from non-living matter, where homeostasis is independent of and thus cannot be reduced to physical mechanisms.

Conundrum three concerned the phenomenon of adaptation, which is a sort of extension of homeostasis where organisms change their form and function to thrive in a particular environment. Different environments make different demands on organisms trying to survive, and the dynamic activity of life is flexible enough to accommodate changes in form and function to adapt to new circumstances. Dr. Turner admits that current science does not have a clear idea about how living entities *know* which form and function will satisfy environmental demands. He suggested that adaptation implies self-knowledge on behalf of the organism regarding what it wants or needs to be.

The conventional view of adaptation claims that the natural selection of genes with optimal characteristics for an environment ("apt function genes") determines the organism's adapted form and function, i.e. that genotype determines phenotype. Instead of this reductionist explanation which ultimately proves to be a tautology, Dr. Turner suggests conceiving adaptation as an expression of homeostasis. He has observed that biological structures are designed consistently and efficiently as a manifestation of homeostasis, like bone structure which exhibits similar design principles to cantilever beams designed by mechanical engineers. Thus, it is reasonable to postulate that mechanisms are a product of adaptation, i.e. that life is fundamental and that mechanisms emerge in that process for the purposes of the living entity.

Adaptive interfaces facilitate the necessary homeostatic regulation for adaptation by acting as a dynamic divider between environments. Dr. Turner explained that the cell membrane is the ultimate adaptive interface since it actively manages the exchange of matter and energy between the extracellular and intracellular environments. It is adaptive because it possesses the capacity to form a cognitive representation of the external environment for the inner homeostatic purposes of the cell. By acknowledging that "life imposes order (not disorder) on environments," Dr. Turner suggested an "extended homeostasis" where homeostasis inside a cell or organism imposes homeostasis in its immediate external environment as well. He explained the gradations of nested hierarchies that can manifest from coalitions of living entities via the principle of "extended homeostasis" from single cells to epithelia, tissues, organ systems, organisms, superorganisms (like the termite colony), populations, ecosystems, up to the biosphere. This principle can lead to the Gaia Theory of James Lovelock and Lynn Margulis.

Conundrum four considered the thermodynamics of life, where planet Earth is an open thermodynamic system where matter and energy flow in a closed loop through adaptive interfaces like different living entities and environments. When matter and energy pass through living entities order is produced (low entropy), and when it leaves them, entropy i.e. disorder increases. The standard model depicts free energy from the sun being absorbed by plants, which are then consumed by primary consumers like small animals, who are consumed by higher-order predators, from which that free solar energy ultimately dissipates as heat. Dr. Turner argued that this is not a good metaphor for considering the thermodynamics of life. He offered a novel wave model instead, depicting low-entropy natural systems like the circulation of the atmosphere and oceans (implying weather and climate) gradually becoming ordered at the peak of a wave which pulls in disorganized free energy from the sun, where living systems sit at the tip of the peak as the most ordered natural phenomena. After being utilized for living processes, matter and energy are eventually released and become disordered, degrading back into equilibrium at the high-entropy base of the wave.

This metaphor for the thermodynamics of life is useful for comprehending how homeostasis and the extended organism fit into an open thermodynamic system. Homeostasis is the maintenance of the shape of the tip of the wave's peak (living entities) despite matter and energy flowing in and out of it, and the extended organism serves to efficiently channel matter and energy through the peak into the tip (living homeostatic systems) in a manner that accommodates a range of environmental circumstances.

The fifth conundrum addressed by Dr. Turner is the cognitive, intentional, and purposeful dimensions of life. Adaptive interfaces are cognitive systems serving as thresholds that intentionally mitigate matter and energy across open thermodynamic systems. Cognition can be understood functionally as the creation of internal mental representations of an external "real world," where the cognitive system typically mediates the mental image to track and conform with the sensory input of the external world. The reverse is also a valid cognitive function, where the external environment can be shaped to conform to a preferred mental image. Dr. Turner suggested that the latter function is the capacity for intentionality and possibly even creativity. He argued that purpose comes into play when the homeostasis of a living entity is imposed on the external environment, which differs from the standard "cybernetic interpretation" due to it not being able to address the origin of goal states in living systems.

The overarching nature of life itself was the topic with which conundrum six was concerned. By comparing cauliflowers and cumulus clouds — identifying their similarities as appearing puffy and white, being open thermodynamic systems, obeying the fourth law of thermodynamics, and having a sort of metabolism and memory — Dr. Turner determined that the difference between them is that a cauliflower *wants* to be a cauliflower. Through living cognitive processes cauliflower maintains its identity for a significant duration of time, while a cumulus cloud does not possess such processes and dissipates relatively shortly after forming.

Turning to philosophical reflection, he discussed Aristotle's idea that an organism is an expression of its nature (Greek: *Bios* [βίος]), where the βίος is the ideal towards which individual organisms strive or the ultimate purpose served by their varying forms and functions. This satisfies the question of the origin of goal states for the cybernetic interpretation of homeostasis. Dr. Turner explained that Aristotle's βίος is the philosophical cousin of Claude Bernard's homeostasis and encouraged scientists to critically consider the "essential purposefulness that sits at the heart of life." He advocated reconsidering Aristotle's βίος as an "organizing philosophical principle of the science of life" and recognized that this gives rise to questions about soul. While such considerations are pertinent to sober and unbiased explanations of the adaptability of organisms, they are apparently incompatible with the idea of evolution which implies change, since βίος does not change.

Accommodating evolution in the framework presented thus far necessitates the heritability of adaptation, which was the seventh conundrum. Tenets of 20th-century biology like the Weismann Barrier and the Central Dogma denounced the idea that adaptations gained throughout an organism's life could be passed to offspring. After the groundbreaking work of Nobel laureate Barbara McClintock, discoveries of 21st-century biology clearly show the inheritance of adaptations to later generations although what guides this remains a mystery. Answering this fundamental question requires clarifying the nature of heritable memory, which surpasses being reduced to a function of genes. Dr. Turner defined heritable memory as any memory that survives beyond the lifetime of the individual organism, thus language, culture, and the termite mounds mentioned previously, all qualify as forms of heritable memory.

The final conundrum addressed by Dr. Turner was the philosophy of life, which he illustrated with a diagram showing three ancient Greek philosophies — Atomism (where order is bottom-up and there is no purpose), Aristotelianism (where order is emergent and purpose is inherent), and Platonism (where order is imposed and purpose is disembodied) — and how these influenced key people in the history of science like Georges Cuvier, Jean-Baptiste Lamarck, Charles Darwin, Claude Bernard, JBS Haldane, and others, leading to the wide array of contemporary approaches such as molecular biology, Neodarwinism, physiology, ontogeny, and the controversial intelligent design (which may be seen as a form of what is popularly thought of as Platonic Idealism). Dr. Turner acknowledged that atomistic philosophy and Neo-Darwinism have served a valuable purpose in the development of modern science, but reiterated advocacy of embracing central tenets of Aristotle's philosophy for progressing toward a coherent theory of life. Any discussion of life must necessarily include mind since physiology and homeostasis are purposeful cognitive phenomena. The presentation concluded that a coherent theory of evolution must account for life's uniqueness and that because evolution is directed by cognitive purpose-laden activity, it is ultimately driven by desire.

What follows was only briefly addressed in the Q&A after Dr. Turner's talk and during the interdisciplinary dialogue at the end of the conference, and is largely inspired by reflection on the conference after it finished.

During his talk at the beginning of the conference, Dr. Denis Noble presented a diagram of an inverted triangle depicting the flow of causation from higher levels of organization to lower ones. The sociotype was the highest level, containing living entities' mental activities like exercising freedom and decision-making, followed by the ecotype or the ecological niche. This corresponded to the highest levels of organization mentioned in Dr. Turner's explanation of nested adaptive interfaces, where both he and Dr. Noble agreed that the ecotype/ecosystem is as alive as the organism. This is Hegel's conception of the "living process."

Actually, Dr. Turner's entire conceptual development from (1) homeostasis in the individual organism, (2) to extended homeostasis through adaptation and the extended organism which are influenced by or integrate aspects of the environment, and (3) the intrinsic nature of life as Aristotle's βίος and the necessity for the inheritance of adaptations across generations, corresponded very nicely to Hegel's development of the three logical levels of life from (1) the "living individual" or the organism as such governed by inner purposiveness, (2) the "living process" or the organism's relation to the environment governed by outer purposiveness, and (3) the "genus process" or the organism's relation to others through reproduction where death of the parents and survival of offspring emphasizes the universal genus intrinsic to and inseparable from the organism.

Recognizing this correspondence and diving deeper into its implications is useful for elucidating the relationship between life and cognition, in addition to conceptualizing life and cognition in a broader systematic philosophical framework that expands on Aristotle's ideas in significant and modern ways. For Hegel, the logical transition from life to cognition occurs once the subjective activity of life is comprehended as a dynamic organic whole, where emphasis is placed on the whole as restless activity. Thus, comprehensive knowledge of the totality of life must accommodate this continuous movement rather than abstracting particular moments of living activity and describing them as static objective phenomena.

Complexity Theory & Purposiveness

Alicia Juarrero, PhD | University of Miami

Latest book: *Context Changes Everything (2023)*

About the Speaker: *President and co-founder of VectorAnalytica, Inc., and Visiting Scholar at the University of Miami. Ongoing research in neurophilosophy is focused on the causal role of context-sensitive constraints in the emergence of mental events such as intentions.*

Abstract: Teleological explanations of natural phenomena rest on a particular understanding of "purposiveness." Specifically, they rest on an 18 th century European understanding of "intrinsic purpose" as self-organization (Juarrero-Roque 1985). This understanding, as Kant noted, in turn rests on a particular view of causality – as recursive and capable of subtending ontically real mereological relationships. This entire framework is closed off to modernity's notion of causes and effects as either (metaphysically) the collision of two independent entities, or (epistemologically) as perceptual correlation between two distinguishable sense perceptions. Lacking real and scientifically respectable notions of emergence and top-down causal relations, philosophy of science generally, and philosophy of mind in particular, have either retreated into epistemological solipsism or proposed a variety of versions of panpsychism. Complex adaptive systems theory, in contrast, conceives of reality as fundamentally dynamic, processual, and based on interactions enabled, constituted, and governed by a variety of constraints (not contradictions). Bottom-up, enabling constraints generate higher-level organizations; top-down, constitutive and governing constraints maintain and preserve the achieved and emergent higher-level organization. They also regulate, modify, and otherwise control those lower-level processes that realize those emergent higher-level organizations.

Such a theory can better illuminate top-down causal relations and therefore purposive or teleological explanations that "look upwards" towards higher-level structures.

Editor's Summary: Dr. Alicia Juarrero began her presentation by summarizing some central tenets of Aristotelean philosophy like the four aspects of cause and two forms of explanation. The formal and final aspects of cause, which were intentionally eliminated from modern science, refer to the inherent essence that makes a thing that particular thing and a thing's goal or purpose (teleology). The material and efficient cause refer to the stuff that constitutes a thing and the external agent that causes it to change, respectively. Dr. Juarrero emphasized that Aristotle held that cause is external to effect such that nothing is the cause of itself, as explained in *Metaphysics* VIII, and that while he accepted the reality of change and development, Aristotle did not embrace the notion of evolution or emergence. His two forms of explanation were *Episteme*, deduction, and *Phronesis*, contextual narrative. Dr. Juarrero explained that modernity (circa 1650) abandoned all forms of cause and explanation except for efficient causation — exertion of physical force like colliding billiard balls — and deduction — which accommodated predictive scientific approaches and discovery of universal laws. This allowed Newtonian mechanics to flourish.

The historical development of this reductionist mechanistic worldview critically influenced the modern view of mereology — the interrelationship between parts and wholes. In the modern view, wholes are reduced to aggregates of parts due to a superficial emphasis on context-independent "primary properties" like mass, while context-dependent properties, an example of which are relational properties like processes, are relegated to an inferior or inconsequential status. This results in dismissing the importance and influence of context altogether which translates to the absence of intentional or purposive explanations in action theory, i.e. top-down intentional activity of a whole influencing the parts is dismissed. This purposeless materialistic action theory cannot distinguish between a wink and a blink, where the former is purposeful and intentional and the latter is reactive and unintentional.

The modern commitment to solely embracing efficient causality has severe limitations. Cartesian Dualism seems to claim that a nonphysical intention serves as an efficient cause that activates the physical body, but Dr. Juarrero objected to this on the basis that it violates causal closure i.e. that all physical events have physical causes. In neuroscience, materialistic action theory seeks a domino effect of neurophysiological processes acting on one another as efficient causes until muscles are activated, but research does not corroborate neurons behaving in this way. Matter understood as passive, inert, unformed material, cannot produce meaningful function and behavior. Muscles are activated when a person engages in intentional behavior like playing sports, but materialistic approaches committed to efficient causation as the only means of explanation cannot deal with this soberly, just as they cannot differentiate between a wink and a blink.

Dr. Juarrero identified the main problem in contemporary action theory as the lack of accommodating recursion and circular causality; this issue traces back to Aristotle's *Metaphysics* VIII. In *Critique of Judgement*, within the "Critique of Teleological Judgment" section, Immanuel Kant recognized this problem. He characterized the phenomenon of a tree producing leaves and the leaves also producing the tree as a "kind of causality unknown to us." Contemporary biologists accept the recursive behavior of autocatalytic chemical reactions, where a product of the reaction also catalyzes the same reaction, but they obscure the causal potency of such phenomena by using terms other than "cause," reserving it for efficient causality. Dr. Juarrero argued that Complex Dynamical Systems Theory provides a solution and can facilitate the reconceptualization of mereological causal relationships through interlevel causality as constraints.

In this regard, constraints do not just restrict systems but also facilitate novel higher-order expression. Constraints (1) are factors or conditions that modulate probable outcomes of other processes/events, (2) regulate or facilitate energy transfer without contributing additional energy, (3) conserve/retain a constant amount of energy themselves, (4) create emergent levels of coherence, and (5) presuppose conditions of open thermodynamic systems far from equilibrium where equilibrium is a state of disorganization. The first kinds of constraints described were context-independent constraints (CICs), which reduce randomness in a system and facilitate system complexity by establishing a "possibility space." Examples include gradients, polarity, charge, and boundary conditions. These alone are not enough to give rise to complex systems. Once CICs are in place, then the second kind of constraint, context-dependent constraints (CDCs), can directly generate complexity by decreasing a system's independence by forming connections between things. These include both spatially-dependent and temporally-dependent constraints. Examples include feedback, catalysts, interfaces, and timing. An instance of timing as a CDC would be the significance of the *moment* that a child playing on a swing set thrusts their legs. *When* they thrust is as important as *how hard* they thrust, in order to swing properly. The timing of the thrust does not itself produce energy, so it cannot be considered an efficient cause, thus this phenomenon does not fit into the contemporary model of causality.

Dr. Juarrero distinguished between two orders of context-dependent constraints, first-order bottom-up "enabling" constraints, and second-order top-down "governing" constraints. Bottom-up CDCs like catalysts and feedback can be called "enabling constraints" since they establish the necessary conditions for complex systems to arise. This idea has been discussed by biologists like Howard Pattee and Stan Salthe who work on hierarchy theory. Bottom-up enabling constraints (1) embed a whole complex system in its context, i.e. prior history and present circumstances, and (2) facilitate interdependence of component parts thus increasing the degree of freedom expressed in the system, which allows complexity to manifest during which novelty is generated through the strong emergence of unique properties — where "strong emergence" refers to emergence that is non-epiphenomenal i.e. the resultant properties are causally potent. The causally potent whole, as a complex dynamical system, is not other than its component parts, i.e. it exhibits circular self-causality as opposed to efficient causality where a cause must be external to and different from its effect.

The closure of constraints for a particular dynamic system establishes a "constraint regime," a dynamic network of constraints including both context-independent and context-dependent constraints that maintain the coherency of a complex system. "Constitutive constraints" maintain the dynamic coherency of the whole, while "governing/regulatory constraints" manage the component parts. The emergent properties of higher levels of organization result from the enhanced degree of freedom enabled by lower-level constraints. A similar idea was also expressed in Dr. Denis Noble's diagram of the inverted triangle depicting successive levels of organization with the highest being the most open/free.

While enabling constraints are considered first-order CDCs due to their function of embedding a complex system in its context thus facilitating its dynamic development, top-down "governing constraints" are understood as second-order CDCs where the whole limits the freedom of its component parts thus constraining the behavior of the parts to the context of the whole. In this way, Dr. Juarrero's theory of interlevel causality as constraints accommodates recursive circular causality. Emphasizing the dynamic role of constraints serves to show the inadequacy of causal powers being limited to efficient causality and also seems to allow for rational consideration of recursive phenomena.

In this framework, action can be described as an intentional behavior influenced by enabling constraints like an individual's past experience and present circumstances. A Complex Dynamical Systems Theory account of intentional causation holds that interdependencies, created through the self-organization and synchronization of neurons, act as enabling constraints that give rise to emergent properties like consciousness and meaning as higher-order governing constraints that in turn produce intentional action. Such intention is what distinguishes a wink from a blink.

For Dr. Juarrero, the difference between a cell, organism, and society is that of the novelty of the constraint regimes, where unique properties emerge at each new level of coherence. She postulated that cells become cancerous when coherency degrades through the weakening of the constraint regime. Speaking in terms of constraints allows the applicability of her theory to extend outside of biology to physics and chemistry since there are many different kinds of top-down control throughout biotic and abiotic phenomena. For instance, dissipative structures like Bénard cells, dust devils, hurricanes, and BZ reactions create metastable order which is what allows them to persist as complex dynamic systems in an environment.

The presentation concluded with a summary of the kind of logic that is appropriate to describe complex dynamical systems. Such descriptions must seriously consider all levels of context (including historical contexts), as well as the first-person perspective of the observer of nature i.e. the scientist. Predictability seems to dwindle in successively higher levels of organization, so descriptions of dynamical systems are usually retrospective accounts of observation. Further, they are often tentative and provisional due to the dynamic nature of context.

What follows was not part of Dr. Juarrero's presentation, but is an elaboration of relevant points addressed by Dr. B. Mādhava Purī, Prof. Brian J Ford, and Dr. Denis Noble, during this session's Q&A and the interdisciplinary dialogue at the end of the conference.

Dr. Juarrero offered a universal framework for considering coherency in natural systems through the role of constraints, which is helpful in significant ways to the progression of modern science, but the source of the constraints is ambiguous. This concern was raised by Dr. B Mādhava Purī. He questioned that if a planet's law of motion is considered a governing constraint, where did this specific constraint come from? How can the movement of a planet emerge from lower-level phenomena where freedom is more restricted? This is unclear.

The distinction between life and non-life is also ambiguous in this framework. While dissipative structures may be self-organizing systems, they are not self-originating. This distinguishes living systems from non-living ones. An organism produces, from within itself, that which it organizes, creating the organic compounds necessary for its subsistence. The initial conditions for an organism are organized within a preexisting parent organism, and after birth, the living processes within the offspring facilitate its development. The consistency with which organisms of particular species are repeatedly produced, millions of times, does not seem to suggest a random act of emergence due to complexification.

The Dynamical Systems Theory account of intentional causation holds that the interdependencies among neurons enable the emergence of consciousness and meaning, but there are several problems with this. The first issue seems to be an internal discrepancy within the theory. If meaning is a higher second-order emergent property of context-dependent constraints, then this implies that lower first-order enabling constraints are meaningless i.e. random. This is problematic because CDCs apparently rely on the "possibility space" opened up by context-independent constraints (CICs), which were defined as reducing randomness in a system to allow for potential complexity. Reduction in system randomness implies an increase in meaning/purpose, which suggests that purpose is intrinsic throughout the entire development of the background of CICs supporting first-order enabling CDCs and the emergence of second-order governing CDCs.

The second issue concerns the discoveries of 21st-century biology, which conclude that all single cells — including neurons — are conscious/cognitive entities that demonstrate environmental awareness and decision-making capacity. If we accept this conclusion, then consciousness (here defined as the capacity for environmental awareness and decision-making, which is more than just top-down control) cannot emerge from the interdependencies among multiple neurons since each individual nerve cell already exhibits conscious behavior. In this regard, Prof. Brian J Ford's comment about bacteria intentionally cooperating while forming structures, where observation of such behavior indicates that they are aware of the benefit of working together, is very important. Based on the videos that Prof. Ford shared, Dr. Denis Noble corroborated that the single-celled organism deploying a pseudopodium to feel out its environment must have had an idea of what it was doing and why it was doing it. In this regard, it is notable that Dr. Noble specifically emphasized downward causation in his account of the levels of organization of living systems.

The ideas presented by Dr. Juarrero clearly provide a rational way for physicists and chemists to cultivate holistic thinking in a manner that does not completely turn their current worldview upside down. It is also clear how these ideas relate to some aspects of biological phenomena. That being said, the notion that life and consciousness are higher-order emergent properties enabled by lower levels of non-living physiochemical phenomena — that the difference between life and non-life is merely a degree of complexity — is problematic, as is the source of constraints. This notion seems to indicate a linear continuity from non-living stages to living stages — albeit one where later stages are not reducible to prior ones — such that life is an irreducible novel quality that emerges from constraints on matter. But we never observe life coming from constrained matter in that way. While Dr. Juarrero clearly articulates that wholes are more than the sum of their parts in that the wholes are causally potent and exercise top-down control, in addition to the bottom-up embedding and enabling function of the parts, her theory still seems to assume that biology is just a complex product of physics and chemistry. That is, biological systems are higher-order systems possessing a greater degree of freedom such that life appears as a novel emergent property of the system, which is ultimately enabled by the interdependencies of lower-order non-living chemical phenomena. While this idea is certainly more advanced than the standard reductionist account of life — due to Dynamical Systems Theory recognizing the synergistic relation among the parts of the biological system, as Kant recognized in the *Critique of Judgement* — there still seems to be reluctance to fully acknowledging the grandeur of life.

The living entity is objectivity (being-in-itself) produced by subjectivity (being-for-itself). Life organizes itself for its own purposes (inner teleology). The objectivity of nonliving entities (material aggregates) is produced by mechanical, chemical, or outer teleological processes, where material aggregates do not have their own purposes — only purposes externally imposed on them. This internal subjectivity of living entities is what allows for sentience, which serves to demarcate the categorical difference between life and non-life. For an individual organism to maintain coherence as a whole, it must internally assimilate what is other from or external to it (digestion of food). This transforms external objectivity, what is in itself, to that which is for the inner subjective necessity of the organism. When we eat something, digestive enzymes break the object down into particulate nutrients which are subjectively utilized by sentient living cells throughout the body to fulfill the cells' desires i.e. carry out their healthy function in blood, bones, tissues, and organs. This is how living bodies are formed and maintained. On the other hand, non-living material aggregates do not require any internal assimilation to form or maintain their bodily/structural constitution.

Sentient life can utilize non-living objects to create artifacts like computers which serve as extensions of life's sentient/cognitive activity, but this does not obscure the stark difference between life and non-life that we feel is so important to explicitly acknowledge. Recognizing the inconceivable uniqueness of life for the precious gift that it is and respecting it as such, without degrading it to appease individuals committed to materialistic ideology, may prove more helpful to the progression of modern science.

10

Conclusion

Aristotle, an ancient Greek philosopher considered a father of modern science to this very day, established a framework for natural science based on four aspects of cause. These were the material, efficient, formal, and final (teleology), where the first two dealt directly with the physical and the last two regarded non-physical influences on the physical. Since the 17th century, however, modern science abandoned the formal and final aspects of cause. Scientists became preoccupied with the material constituents of natural phenomena and efficient causality, exemplified by the collision of billiard balls where the exchange of physical force between objects carries an exchange of energy. The historic development of scientific thought that ensued produced a materialistic reductionist science that viewed living entities as purposeless mechanical conglomerates and consciousness as inconsequential.

Throughout the 20th century, the notion that the form and function of organisms were determined by genes became prevalent. This influenced society's worldview to believe that all personal qualities and behaviors are a result of genetic determinism, as opposed to free will, conscious decision-making, and contextual circumstances. The gene-centric worldview — exemplified by sayings like "we are born selfish" and "it's in their DNA" espoused by scientists like Richard Dawkins — has negatively impacted society and influenced ideas in a wide array of other fields from economics to philosophy. At the turn of the century, the Human Genome Project (HGP) — which ran from 1990 to 2003 and identified, mapped, and sequenced every gene in the human genome — was the culmination of faith in genetic determinism. HGP promised to revolutionize biology and medicine at its outset, but this was based on an overestimation of the causal potency of genes in the organism. Over two decades after the project's completion, scientists still lament how it fell short of those big promises.

Biologists are realizing that genes are just chemicals whose function is determined by higher-order living processes in cells and organisms. Thus, 21st-century biology is becoming more of an organismic-centered enterprise that views the function of lifeless genes in the broader context of the living entity. This shift in biology has necessarily given rise to a change in language that is reminiscent of Aristotle's original teleological paradigm. Living processes both internal and external to the organism are goal-directed, which require consideration of purpose, desire, intention, and cognition to adequately explain. Astounding evidence shows that consciousness is ubiquitous from the smallest single cell throughout multicellular organisms like plants and animals; this frustrates attempts to explain consciousness as an epiphenomenon or emergent property of interdependent firing neurons since even a single neuron (nerve cell) exhibits conscious behavior like decision-making and environmental awareness.

This revolution in biology has exposed the shortcomings of reductionist mechanistic science's ability to provide an accurate and comprehensive account of living phenomena. All that said, the seriousness of society's conditioning to understand life in mechanistic terms is not being belittled. This outdated notion still permeates textbooks throughout most schools, where the complex cognitive behavior of living cells is almost entirely ignored and cells themselves are depicted as grossly oversimplified globules. Such misrepresentations allow scientists to entertain notions that accurate computer models of cells are within their reach, for which there are large sums of grant funding. However, this is an illusion.

As the reality of the shortcomings of reductionist mechanistic ideology are embraced by modern philosophers and scientists alike, more holistic thinking is becoming prevalent as seen in systems theories. Dynamical systems theory recognizes the irreducible complexity of self-organized abiotic and biotic systems like dissipative structures and organisms respectively, and attempts to provide conceptual frameworks to rationally account for the cause and effect of such complexity. The field of systems biology recognizes that at each level of biological activity, individual components are integrated into a coherent network, and understanding the logic of how each system operates requires acknowledging that the function of the whole is goal-directed and greater than the sum of its parts. A central conceptual theme throughout systems thinking is the recognition of different levels of organization and the ability to tailor one's questioning to the appropriate level. As history shows, failure to do this results in reductionistic fallacies. When considering these levels of organization and the causal relationship between them, there seem to be two perspectives that arise. The first is upward causation — where lower levels give rise to higher levels (bottom-up) — which includes but is not limited to the genetic determinism of organisms. The second is downward causation, where higher levels produce and influence lower ones (top-down). Causal relations between higher- and lower-order levels of organization surpass the limitation of just efficient causality to include all four aspects enunciated by Aristotle, including non-physical influences on the physical. Some systems philosophers and scientists even suggest that there are a couple more. These recent developments have revitalized interest in mereology — the interrelationship between parts and wholes — and prove the value of philosophy for scientific advancement.

Novel approaches among insightful contemporary physiologists show how living phenomena like homeostasis — comprehended as an irreducible fundamental non-physical activity of living entities that maintain their individual coherence — practically facilitate the dynamic relationships between different levels of organization in nature. When the internal activity of homeostasis is imposed onto a living entity's immediate external environment, such as when a colony of termites creates a massive above-ground mound that serves as a wind-driven lung for the colony living deep underground, then otherwise inert lifeless matter like dirt becomes as alive as the organisms themselves, forming an "extended organism." The phenomenon sometimes called "extended homeostasis" gives rise to gradations of nested hierarchies of coalitions of living entities — from single cells to epithelia, tissues, organ systems, organisms, superorganisms (like a colony of termites or bees), populations, ecosystems, up to the biosphere. The ideas of extended homeostasis and extended organisms have further implications for the Gaia Theory.

Such phenomena apparently exhibit upward (bottom-up) causality, but when the initial conditions of environmental pressure that constrain or stress an individual cell or organism (prompting them to cooperate with others to relieve the stress) is accounted for, then recognition of a more circular causality becomes evident. Homeostasis, as it has been explained above, can be conceived as a conceptual descendent of what Aristotle called *Bios* (βίος), the intrinsic purposiveness sitting at the heart of life that is served by an organism's various forms and functions. Homeostasis and βίος denote a fundamental organizing principle central to the science of life. Despite doctrines from 20th-century gene-centric biology rejecting the idea of heritable adaptations, the pioneering work of Nobel laureate Barbara McClintock allowed 21st-century biology to make discoveries showing that adaptations gained throughout the course of a parent's lifetime were inherited by later generations. Heritable memory surpasses being reduced to a function of genes since any memory that survives beyond the lifetime of an individual organism like language, culture, and the termite mounds mentioned previously, all qualify as forms of heritable memory.

These collective phenomena of the homeostatic nature of individual living entities maintaining coherency by adapting themself, extending through and adapting their environment, and passing such adaptations on to offspring (the inheritance of both internal phenotypic changes and reformed external structures) which ultimately serves to maintain the coherency of the species, emphasizes (1) the organism as such governed by inner purposiveness, (2) the organism's relation to the environment governed by outer purposiveness, and (3) the organism's relation to others through reproduction where death of the parents and survival of offspring emphasizes a universal aspect intrinsic to and inseparable from the organisms. This conceptual development supported by empirical observation is strikingly similar to what G.W.F. Hegel presented around 1816 in his *Science of Logic* in the development from the "living individual" through the "living process," to the "genus process." Acknowledging this correspondence and exploring its implications serves to conceptualize life and cognition in a more comprehensive systematic philosophical framework that expands on Aristotle's ideas in important and relevant ways. Hegel said that the logical transition from life to cognition happens once the subjective activity of life is comprehended as a dynamic organic whole, emphasizing the whole as restless activity. A truly coherent conception of life must accommodate this continuous movement as opposed to abstracting particular moments of living activity and understanding them as reified static objective phenomena.

For the majority of contemporary scientists and philosophers to seriously consider what is being presented here would require the adoption of a humble, reflective, and self-corrective attitude. To this end, the integration of a kind of yogic meditation would be beneficial for scientists, since meditation helps to cultivate self-consciousness. A self-conscious scientist is interested in being keenly aware of the contribution that the self makes to the empirical observation of natural phenomena, thus they might consider:

- What axioms are employed in scientific experimentation?
- What presumptions shape the boundary conditions for particular experiments?
- What biases influence discrimination between noise and signal in data collection?
- Is there a necessity for such axioms, presumptions, and biases?

Acknowledging the contribution of the self in the production of empiric knowledge may disillusion the belief that scientists occupy a third-person perspective as passive observers of nature. An observer's subjective consciousness is entangled with the object(s) of observation — a cow is sacred for some and a meal for others. Formulating a scientific method that recognizes this fact and accommodates a systematic first-person phenomenological account of how purely logical thought determinations serve as a foundation for further determined sensuous thoughts about natural phenomena would integrate empirical (external observation) and meditative (internal observation) approaches, thus widening the epistemological scope through which modern science can discover the truth.

Those who are skeptical about first-person perspectives of consciousness being legitimately integrated into a systematic science of observation and experiment may find solace in contemplating that so-called passive third-person observation occurs within the first-person experience of a scientist, who say "*I* see," "*I* hear," "*I* observe," and "*I* think." So, the fact that scientists can reach mutual agreement about natural phenomena means that certain aspects of first-person experience are not merely idiosyncratic, but are common among individuals. As Hegel explains in his *Phenomenology of Spirit*, consciousness denotes an interdependent relationship between subject and object where each dialectically determines the other — but this does not mean that each individual simply lives in their own world of subjectively determined objects. Individual subjects do not exist in isolation independent from others, we are all intimately and necessarily interdependent on each other. We participate in a shared objective experience while simultaneously maintaining our individual subjective perspective. The degree to which individual subjects comprehend the universal aspect of life — the highest level of organization within which all differentiated activity occurs — determines the degree of harmony that we experience in our exchanges with others, which further determines the (peaceful or chaotic) state of the shared world we all live in.

The Vedāntic view offers a comprehensive holistic framework — which has remained relevant for over five millennia — for conceiving life and cognition at the intersection of science, philosophy, and religion. Vedic knowledge is sensitive to the nuances of downward (top-down) causality, as seen in descriptions of the emanation of the material domain from the spiritual, which starts with eternally conscious and blissful inifinite Spirit (God, Divine Personality), to which eternally conscious and blissful finite spirit (individual souls) is a marginal potency, where the souls may choose to engage in different kinds of reciprocal loving service activities with God through the internal potency, or choose to be self-centered enjoyers subjected to illusory experience through the external potency where the further emanation of functionally defined subtle material elements (intelligence, false-ego, and mind) and gross material elements (ether, air, fire, water, and earth) constitute the worldly experience of birth and death. The idea that higher levels of organization possess higher degrees of freedom and that lower levels are further restricted is evident in this ancient framework, where the spiritual domain entails eternal life and the material world is subjected to the dichotomy of life and death.

Vedic epistemology describes methods for attaining both material knowledge and spiritual Truth, where such methods gradually suggest that the finite knower lessens their attempt to forcefully take knowledge and increases cultivation of humble receptivity to that which is beyond the assertive grasp of the finite individual. Such methods include familiar bottom-up (ascending) approaches like direct sense apprehension and learning from the sensuous apprehension of others, which characterize modern science, in addition to mystical knowledge gained through yogic meditation. These are considered inferior due to the irrationality of the finite trying to claim the infinite for its own self-centered enjoyment. Vedāntic authorities like the Vaiṣṇavas conclusively recommend descending (top-down) approaches to realizing spiritual Truth, which require self-surrender to the guidance of a bonafide spiritual preceptor (Guru) who serves as a philsophical midwife for the surrendered soul.

The application of Hegel's philosophy to biology, especially systems biology, is an emerging field — from 2009 through 2023 biologists and philosophers from France, Germany, Japan, Portugal, Netherlands, and Estonia recognized and explored this topic. We believe this field will make a tremendous impact on contemporary science. These Science & Scientist 2023 conference proceedings were written to summarize the talks and humbly present the common thread of themes from the diverse perspectives that are relevant to topics addressed by Hegel. After reflecting on the conference, we were surprised by the significant degree to which these themes dealing with systems thinking correspond to Hegelian philosophy. We were elated to find that some work has been published in this connection over the past 15 years. The Princeton Bhakti Vedanta Institute hopes to make work in this emerging field more visible and accessible to other scientists and philosophers — especially from the USA, UK, and India — in addition to contributing to this field ourselves through facilitating deeper contemplation via our conferences and publications.

Biologists from Estonia and France recognize the historical development from Claude Bernard's 19th-century physiology, to 20th-century reductionist molecular biology, and now the 21st century's more holistic field of systems biology, as following the Hegelian scheme of the progress of history from thesis to antithesis then synthesis. [1] A German developmental biologist recognizes Hegel's (1770-1831) dialectics as "the first systematic system theory," preceding even Bernard (1813-1878), and studies embryonic induction in light of Hegel's philosophy. [2] A Japanese science professor recognizes Hegel presenting the theme of self-organization in the sense of Ilya Prigogine (1917-2003) and Stuart Kauffman's (1939-) work and credited Hegel with insights ahead of his time. [3] In what has become known as the organizational account (OA) — the origins of which contemporary scholars trace to Kant, Bernard, and Jean Piaget — living entities are recognized as a very particular kind of self-organizing system whose fundamental properties are "inherently related to self-determination." [4] Since self-determination is a subject with which Hegel was greatly concerned, using it to practically define freedom, it is reasonable that a Portugese philosopher is interested in both how Hegel's philosophy develops important OA ideas and how OA ideas may assist in clarifying some of Hegel's thought. [5]

A doctoral student from Durham University dedicated his thesis to strengthening the connection between Hegelian dialectics and the organizational approach as an alternative to the modern synthesis, contributing to the development of the Extended Evolutionary Synthesis. [6] This inspired a philosophy professor from the Netherlands — who has published on viewing technoscience through the lens of dialectics and phenomenology [7] — to present some of the doctoral thesis' ideas in *Dialectical Systems: A Forum In Biology, Ecology And Cognitive Science*, conceiving dialectic causality as reciprocal interaction that gives rise to individuality, where individuality is not given at the start but is a result of the interaction. [8]

The implications of dialectic causality for describing the dynamic between organism and environment are profound, as seen in the Hegelian account of the nutritive process where an organism turns to the external world, takes a foreign substance into itself, and overcomes this otherness through assimilating the external element into an essential self-sustaining substance, thus returning to itself.

Through digestion and assimilation, the organism aims at realizing its own identity. In nutrition the organism does not lose its distinctiveness, nor does it dissolve into the world, but it rather imposes its own determination on what it eats.

This philosophical model, by which metabolic identity is thought of as dialectical, has been quite robust. Hans Jonas, for example, has devoted this understanding of metabolism to the constitution of an interiority, autonomy, and self-identity of the organism. In his 1966 book, the *Phenomenon of Life*, Jonas holds that it is in the dialectical relationship that organisms establish with their environment through metabolism that they realize their identity [...] In this sense, the organism must be understood as the result of its constant metabolizing activity. [9]

Hegel's conceptual framework not only accounts for the logical dimension underlying nature but through the necessity of reason develops from nature into Spirit, where freedom is most fully expressed. He clearly articulates that "Spirit is the truth of Nature" [10] and that "Freedom is the truth of Spirit." [11] We cannot predict where embracing this philosophical framework will lead scientific discovery, however, due to its sheer comprehensiveness — which rationally accommodates contradiction through dialectics and a science of the experience of consciousness as phenomenology — it can serve as an unrivaled foundation for holistic scientific progress in the 21st century. For further reading on the relationship between Aristotle's and Hegel's philosophy we recommend *Hegel and Aristotle* [12], and to understand the nuances of Hegelian phenomenology compared to other views like Husserl, *Hegel and Phenomenology* [13] is suggested.

This conference was intended to serve as an open-minded platform for the exchange of ideas pertinent to the positive progression of contemporary scientific, philosophical, and religious research. We humbly hope that these proceedings may be of some value in this regard. All relevant conference videos and presentation materials from the speakers can be found at www.bviscs.org/ss23.

REFERENCES

1. Saks, Valdur, Claire Monge, and Rita Guzun. 2009. "Philosophical Basis and Some Historical Aspects of Systems Biology: From Hegel to Noble - Applications for Bioenergetic Research." Int. J. Mol. Sci. 10: 1161-1192. https://doi.org/10.3390/ijms10031161

2. Niehrs, Christof. 2011. "Dialectics, systems biology and embryonic induction." *Differentiation* 81(4): 209-216. https://doi.org/10.1016/j.diff.2010.10.004

3. Takahashi, Kazuyuki Ikko. 2015. "Complex Systems Biology and Hegel's Philosophy." *Proceedings Of The 59th Annual Meeting Of The International Society For The Systems Sciences* 1:1. https://journals.isss.org/index.php/proceedings59th/article/view/2658

4. Corti, Luca. 2022. "The 'Is' and the 'Ought' of the Animal Organism: Hegel's Account of Biological Normativity." *History and Philosophy of the Life Sciences* 44(2): 17. https://doi.org/10.1007/s40656-022-00498-8

5. Ibid.

REFERENCES

6. Sandnes-Haukedal, Rasmus. 2023. "Agency and Organisation: The Dialectics of Nature and Life." Doctoral thesis, Durham University. http://etheses.dur.ac.uk/14893/

7. Zwart, Hub. 2022. *Continental Philosophy of Technoscience*. Springer. https://doi.org/10.1007/978-3-030-84570-4

8. ———. 2023. "The Organism As A Subject: Hegel On Nature, Subjectivity, And Interconnectedness." *Dialectical Systems*. https://www.dialecticalsystems.eu/contributions/the-organism-as-a-subject-hegel-on-nature-subjectivity-and-interconnectedness/

9. Bognon-Küss, Cécilia. 2023. "Metabolism in Crisis? A New Interplay Between Physiology and Ecology" in *Vitalism and Its Legacy in Twentieth Century Life Sciences and Philosophy* by Christopher Donohue and Charles T. Wolfe. Springer. https://doi.org/10.1007/978-3-031-12604-8

10. Hegel, G.W.F. 1894. *Hegel's Philosophy of [Spirit]* §388. Clarendon Press. https://www.gutenberg.org/files/39064/39064-h/39064-h.html#toc5

11. ———. 1837. *Hegel's Philosophy of History*, General Introduction 1. https://www.marxists.org/reference/archive/hegel/works/hi/introduction.htm

12. Ferrarin, Alfredo. 2001. *Hegel and Aristotle*. Cambridge University Press. https://doi.org/10.1017/CBO9780511498107

13. ———, Dermot Moran, Elisa Magrì, et. al. 2019. *Hegel and Phenomenology*. Springer. https://doi.org/10.1007/978-3-030-17546-7

BVISCS PEER-REVIEWED PUBLICATIONS

Muni, B. V. (2019). "Quantum Mechanics Shows the Limit of Naïve Realism" in *Quantum Reality and Theory of Śūnya*, 249-269. Siddheshwar Rameshwar Bhatt (ed.). Springer Singapore. doi.org/10.1007/978-981-13-1957-0

Shanta, B. N. (2019). "Subjective Evolution of Consciousness in Modern Science and Vedāntic Philosophy: Particulate Concept to Quantum Mechanics in Modern Science and Śūnyavāda to Acintya-Bhedābheda-Tattva in Vedānta" in *Quantum Reality and Theory of Śūnya*, 271-282. Siddheshwar Rameshwar Bhatt (ed.). Springer Singapore. doi.org/10.1007/978-981-13-1957-0

Shanta, B. N. (2018). "21st century biology establishes that life is beyond chemicals." *International Journal of Recent Trends in Science And Technology*, P-ISSN 2277-2812 E-ISSN 2249-8109 Special Issue, ACAEE: 2018 pp 311-329

Shanta, B. N., Muni, B. V. (2016). "Why Biology is Beyond Physical Sciences?" Advances in Life Sciences, 6 (1): 13-30. doi:10.5923/j.als.20160601.03

Shanta, B. N. (2016). "Vedāntic view of life: Reply to Gustavo Caetano-Anollés." *Communicative & Integrative Biology*, 9(2), e1160191. doi.org/10.1080/19420889.2016.1160191.

Shanta, B. N. (2015). "Life and consciousness – The Vedāntic view." *Communicative & Integrative Biology*, 8(5), e1085138. doi.org/10.1080/19420889.2015.1085138.

Shanta, B. N. (2014). "The Chronology of Geological Column: An Incomplete Tool to Search Georesources." In *Geo-Resources*, 609-625. K. L. Shrivastava, Arun Kumar, P. K. Srivastav, and H. P. Srivastava. Jodhpur, Rajasthan, India: Scientific Publishers. doi: 10.13140/RG.2.1.4409.4808

www.ingramcontent.com/pod-product-compliance
Lightning Source LLC
Chambersburg PA
CBHW072215070526
44585CB00015B/1349